STUDENT'S SOLUTIONS MANUAL

to accompany

UNIVERSITY PHYSICS

STUDENT'S SOLUTIONS MANUAL

to accompany

UNIVERSITY PHYSICS

GEORGE ARFKEN

DAVID GRIFFING

DONALD C. KELLY

JOSEPH PRIEST

Miami University
Oxford, Ohio

ACADEMIC PRESS, INC.

(Harcourt Brace Jovanovich, Publishers)
Orlando San Diego San Francisco New York London
Toronto Montreal Sydney Tokyo São Paulo

Cover art from Isaac Newton's *Principia, Volume 2,* page 551, 1966.
Reprinted by permission of the University of California Press.

Academic Press, Inc.
Orlando, Florida 32887

United Kingdom Edition published by
Academic Press, Inc. (London) Ltd.
24/28 Oval Road, London NW1 7DX

ISBN: 0-12-059867-1

Printed in the United States of America

TABLE OF CONTENTS

SOLUTIONS/HINTS/PROGRAMS

This manual contains solutions or hints for over 450 of the problems in University Physics. Solutions are given for problems that illustrate points not covered in text Examples and for problems that extend ideas developed in the text. Hints are supplied for some of the more challenging problems and in instances where the hint can shorten the solution.

The statement of the problem is not repeated and we have purposely kept the treatment brief - trusting that you will learn to read between the lines.

VECTORS are indicated by a wavy underline, e.g. $\utilde{v}$, $\utilde{A}$, $\utilde{i}$

A broad arrow $\Longrightarrow$ means "implies" or "leads to".

In order to solve a physics problem you need to know what principle(s) to apply and how to formulate them mathematically. Because the problems are grouped by Section it is usually clear what principles are to be applied. Formulating the problem - "setting it up" - is what you may find difficult. The Problem Solving Guide introduced in Chapter 3 will help you develop a systematic approach. Ultimately, there is just one remedy - experience. You must solve problems until the patterns become familiar.

Following the solutions/hints is a collection of computer programs written in the BASIC language. These programs allow you to explore certain problems, and the related physical ideas, in greater depth. Program listings are given for three popular microcomputers, Radio Shack, IBM, and Apple.

Physics is fun. Enjoy it.

1.7 a) (19 yr)·(3.16×10^7 s/yr) = 6.00×10^8 s

b) An atomic clock accurate to one part in 10^{12} could measure 6×10^8 s to within $6 \times 10^8 \text{s}/10^{12} = 6 \times 10^{-4}$ s.

1.12 a) Assume that a product of the given quantities raised to unknown powers has the dimension of length.

$$[c^x h^y m_p^z] = L$$

Substituting the dimensions for c, h, and m_p gives

$$L^{x+2y} M^{y+z} T^{-x-y} = L^1 M^0 T^0$$

Because L, M, and T are fundamental quantities they are independent. This gives 3 equations for the 3 exponents: $x+2y = 1$; $y+z = 0$; $-x-y = 0$. These are solved by $x = -1$, $y = 1$, $z = -1$. Thus $[h/m_p c] = L$

b) Use the same technique to prove that $[h/m_p c^2] = T$

1.13, 1.14 HINT: Use the technique developed in the solution of 1.12. In 1.13 this means taking $[a^p x^q] = [v] = LT^{-1}$, where p and q are the unknowns to be determined. In 1.14 you set $[v^p r^q] = [a] = LT^{-2}$.

1.22 b) The basic idea in using conversion factors is this:

ALL CONVERSION FACTORS EQUAL UNITY

(Intrinsically, but not numerically)

For example, the conversion factor to go from inches to feet is 1 ft/ 12 in. Because 1 ft = 12 in, the conversion factor is the ratio of two equal quantities. Thus the conversion factor is intrinsically equal to unity although numerically it equals 12. The "trick" is to multiply the given quantity by a series (perhaps) of conversion factors which eliminate "undesired" units in favor of the desired units. In 1.22b) we start with m/s and work toward ft/ns.

$$3.00 \times 10^8 \text{ m/s}(1 \text{ s}/10^9 \text{ ns})(100 \text{ cm}/1 \text{ m})(1 \text{ in}/2.54 \text{ cm})(1 \text{ ft}/12 \text{ in})$$
$$= 0.984 \text{ ft/ns}$$

2.3 Analytic Solution:

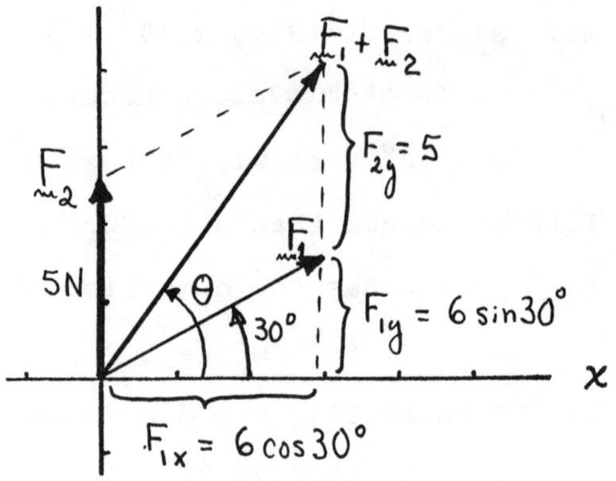

$\mathbf{F}_2 = 5\mathbf{j}$ N

$\mathbf{F}_1 = 6\cos 30^{\circ}\mathbf{i} + 6\sin 30^{\circ}\mathbf{j}$ N

$\mathbf{F}_1 = 5.20\mathbf{i} + 3.00\mathbf{j}$ N

$\mathbf{F}_1 + \mathbf{F}_2 = 5.20\mathbf{i} + 8.00\mathbf{j}$ N

$|\mathbf{F}_1 + \mathbf{F}_2| = [(5.20)^2 + (8.00)^2]^{\frac{1}{2}}$ N

$|\mathbf{F}_1 + \mathbf{F}_2| = 9.54$ N

$\theta = \arctan(8.00/5.20) = 57.0^{\circ}$

2.7 HINT: In relative velocity problems the basic relation has the form (A, B, and C denote arbitrary objects or reference frames)

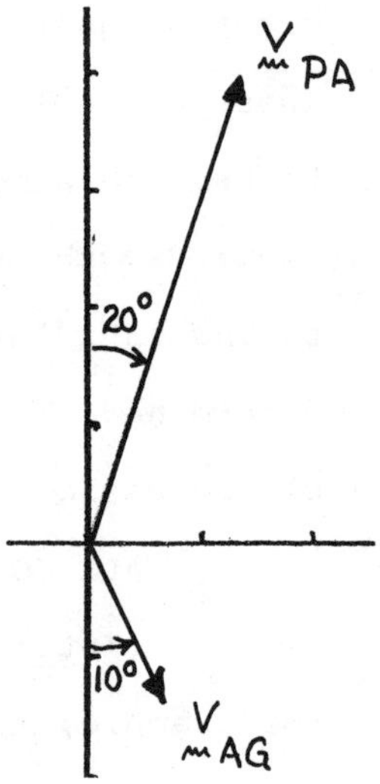

$$\mathbf{v}_{AC} = \mathbf{v}_{AB} + \mathbf{v}_{BC}$$

where $\mathbf{v}_{AC}$ = velocity of A relative to C

$\mathbf{v}_{AB}$ = velocity of A relative to B

$\mathbf{v}_{BC}$ = velocity of B relative to C

In this problem (P ⇒ plane, G ⇒ ground, A ⇒ air)

$$\mathbf{v}_{PG} = \mathbf{v}_{PA} + \mathbf{v}_{AG}$$

The figure shows $\mathbf{v}_{PA}$ and $\mathbf{v}_{AG}$. You can work out the x- and y-components of $\mathbf{v}_{PA}$ and $\mathbf{v}_{AG}$ and then perform the vector addition by components to obtain the components of $\mathbf{v}_{PG}$.

2.12 The sketch shows that the angle between the weight vector and the plane is 55°. The component of the weight along the plane is thus

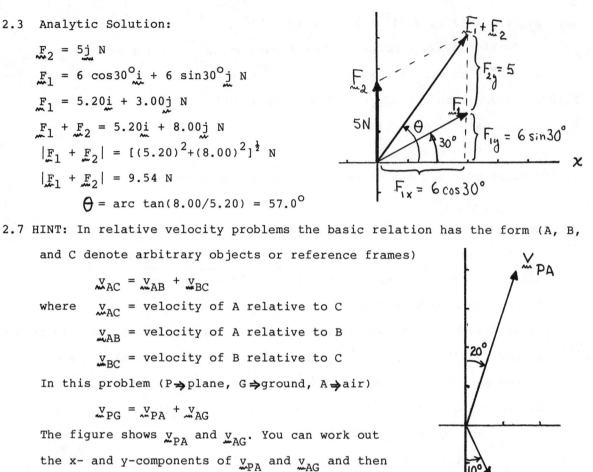

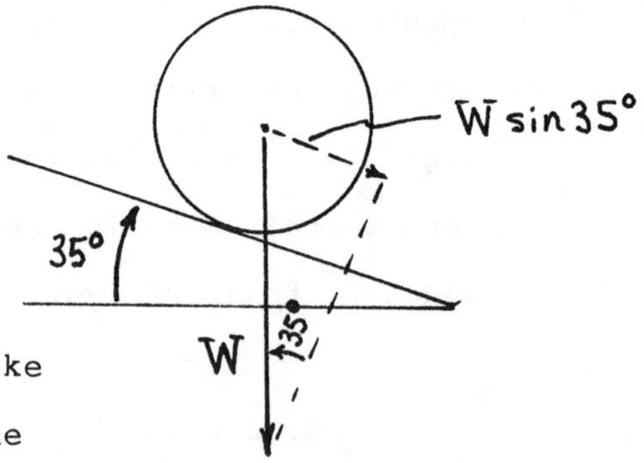

$W\sin 35^{\circ} = 450(0.573) = 258$ N

REMARK: When making sketches for problems like this where you must find components, make the angle look distinctly different from 45°, regardless of the actual value of the angle. This makes it less likely that you will make a mistake and choose the cosine of the angle when the sine is needed, or vice versa.

2.16 HINT: With $\mathbf{v}_2$ = velocity of boat relative to water (1st hour) and $\mathbf{v}_1$ = velocity of water relative to land

$\mathbf{v}_2 + \mathbf{v}_1$ = velocity of boat relative to land (1st hour)

During the second hour the velocity of the boat relative to the water is reversed and so is simply $-\mathbf{v}_2$. Thus, the velocity of the boat relative to the land during the second hour is $-\mathbf{v}_2 + \mathbf{v}_1$. The displacement at the end of 2 hr is given by

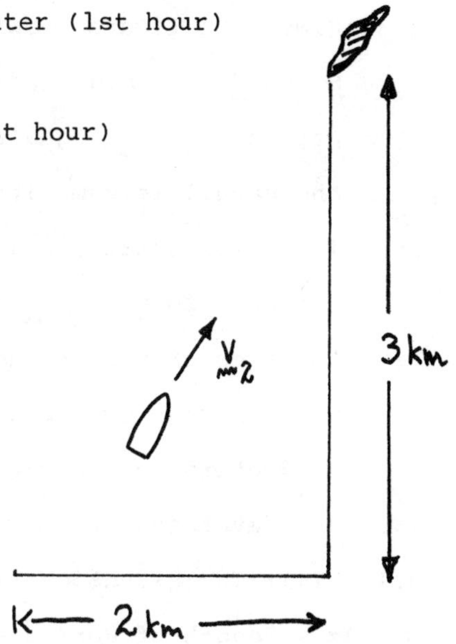

displacement = $\mathbf{v}$(1st hr)·1 hr + $\mathbf{v}$(2nd hr)·1 hr

The fact that the boat is east of the island at the end of the first hour lets you determine b. The geometry of the figure allows you to determine the ratio a/b.

2.18 $\mathbf{r} = \mathbf{i}x + \mathbf{j}y + \mathbf{k}z$ where $x = r\cos\alpha$; $y = r\cos\beta$; $z = r\cos\gamma$

Substitute these expressions into $r^2 = x^2 + y^2 + z^2$

$r^2 = r^2\cos^2\alpha + r^2\cos^2\beta + r^2\cos^2\gamma$

Dividing by r^2 establishes the desired result

$1 = \cos^2\alpha + \cos^2\beta + \cos^2\gamma$

2.24 HINT: a) Use $\mathbf{A}\cdot\mathbf{B} = AB\cos\theta$

b) Use $\mathbf{A}\cdot\mathbf{B} = A_xB_x + A_yB_y$

2.26 HINT: Make use of the two different forms for the dot product. Equating these two expressions for $\mathbf{F}\cdot\mathbf{G}$ allows you to calculate the cosine of the angle between the vectors and then the angle itself.

2.29 From the figure $\mathbf{C} = \mathbf{A} + \mathbf{B}$

Form the dot product $\mathbf{C}\cdot\mathbf{C} = C^2$

$C^2 = \mathbf{C}\cdot\mathbf{C} = (\mathbf{A} + \mathbf{B})\cdot(\mathbf{A} + \mathbf{B})$

$C^2 = \mathbf{A}\cdot\mathbf{A} + \mathbf{B}\cdot\mathbf{B} + 2\mathbf{A}\cdot\mathbf{B}$

$C^2 = A^2 + B^2 + 2AB\cos\theta$

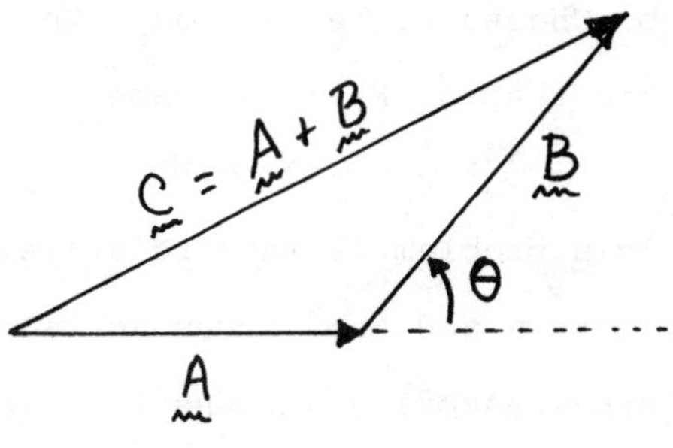

2.34 HINT: The figure shows two vectors, **S** and **T**. The magnitude of their cross product $|\mathbf{S} \times \mathbf{T}|$ equals the area of the parallelogram with sides **S** and **T**. It follows that $\frac{1}{2}|\mathbf{S} \times \mathbf{T}|$ is the area of the triangle formed by **S** and **T**. To determine the area of the triangle ABC in 2.14 you must express two "sides" of the triangle as vectors, compute their cross product, and determine its magnitude.

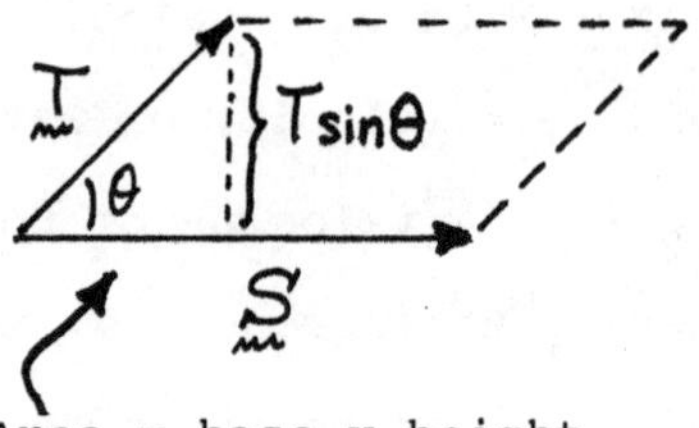

Area = base x height

$= ST\sin\theta = |\mathbf{S} \times \mathbf{T}|$

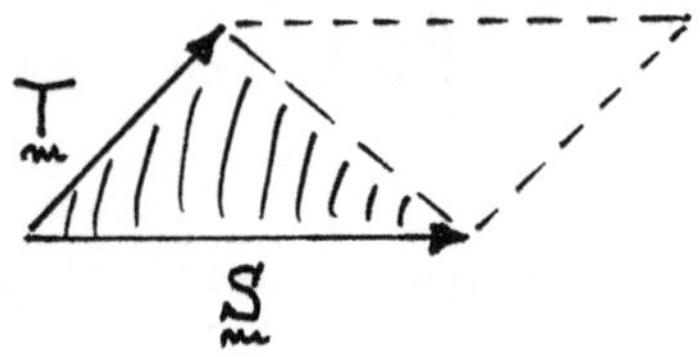

3.1 HINT: The extension (x) of the spring is given by

x = length of spring - 10 cm

3.4 Let's start by guessing. With the fish in equilibrium its weight must be balanced by an equal force F exerted by the rope, so we expect F = W = 4000 N. To guess T we can imagine that there is an identical fish of weight W hanging on the left rope. That would maintain equilibrium and it lets us see that there is a total downward force of 2W plus the pulley weight. This gives a total of 2(4000)+100 = 8100 N being supported by the pulley cable. So, we expect T = 8100 N. Let's follow the Problem Solving Guide to verify these guesses.

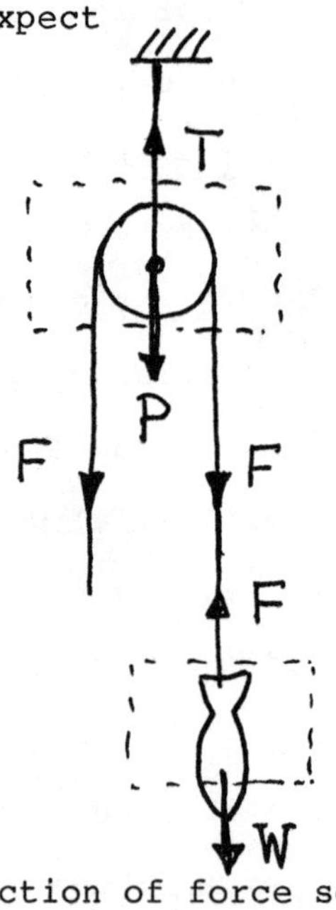

a) Choose fish as body. First condition gives

$$\sum F = F - W = 0 \Rightarrow F = W = 4000 \text{ N}$$

b) Choose pulley as body. Pulley acts only to change direction of force so the tension F is the same on both sides. First condition gives

$$\sum F = T - 2F - P = 0 \Rightarrow T = 2F + P = 2(4000) + 100 = 8100 \text{ N}$$

This problem is made relatively easy because in each part there is just one unknown and the forces act along one direction - there are no components, or simultaneous equations that make the problem algebraically complicated.

3.7 HINT: Choose the body to be the girl plus the weightless seat. There are five forces acting on the body. Explain why the four upward forces are all equal and apply the 1st condition of equilibrium.

3.8

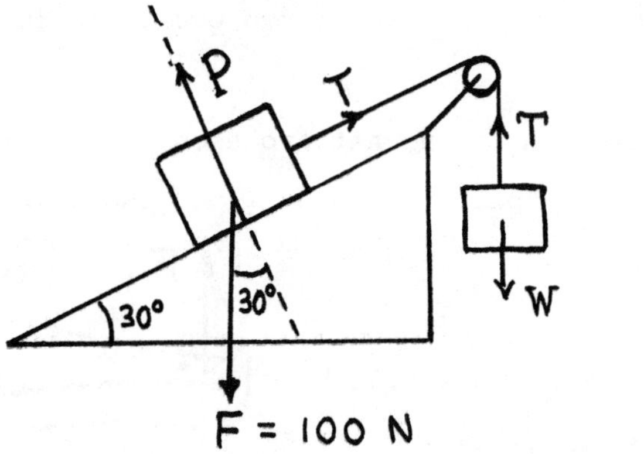

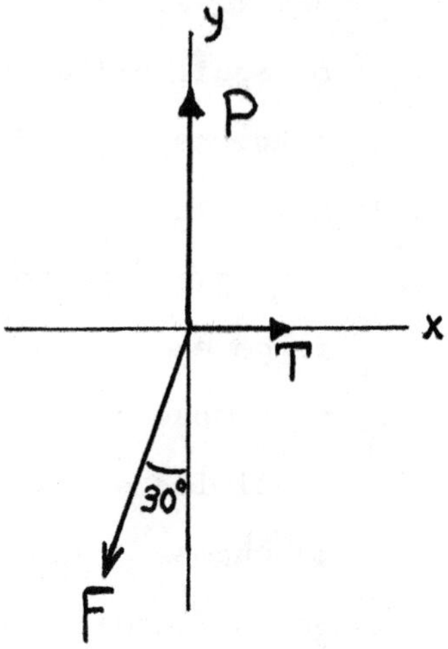

a) $F_{parallel} = F\sin 30^\circ = 100(0.500) = 50$ N

$F_{perp} = F\cos 30^\circ = 100(0.866) = 86.6$ N

b) $\sum F_x = T - F_{parallel} = 0 \Rightarrow T = 50$ N

$\sum F_y = P - F_{perp} = 0 \Rightarrow P = F_{perp} = 86.6$ N

c) Choose hanging weight as body. $\sum F = T - W = 0 \Rightarrow W = T = 50$ N

3.13

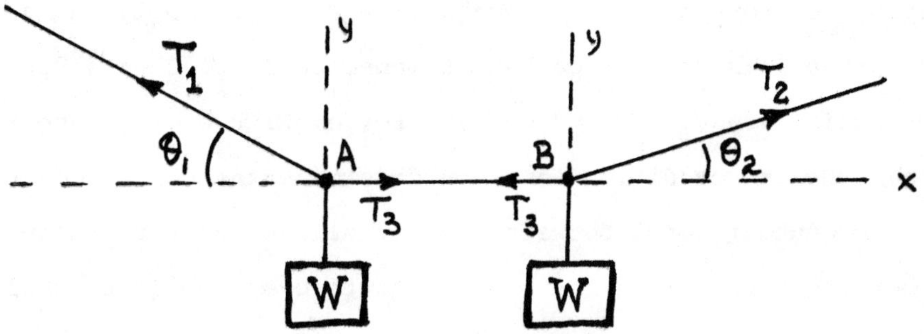

a) Choose the tie points A and B as bodies and apply $\sum F_y = 0$. This gives

At A $T_1\sin\theta_1 - W = 0$; At B $T_2\sin\theta_2 - W = 0$

from which we conclude $T_1\sin\theta_1 = T_2\sin\theta_2$.

If $\theta_1 = \theta_2$ then $T_1 = T_2$.

b) From a) we have $T_2 = T_1 = W/\sin\theta_1 = 200/\sin 8^\circ = 1440$ N

Apply $\sum F_x = 0$ at A to get

$T_3 - T_1\cos\theta_1 = 0 \Rightarrow T_3 = T_1\cos\theta_1 = 1440\cos 8^\circ = 1420$ N

3.15 HINT: a) $\theta_1 = \theta_2 = 30^\circ$ follows from part b) where $\theta_1 + \theta_2 = 60^\circ$. The thrust forces are perpendicular to the walls so the directions of T_1 and T_2 are known. Choose the bottle as a body and apply the first condition of equilibrium in component form. This gives two equations in the two unknowns, T_1 and T_2.

3.21 HINT: The rope is in equilibrium so the net froce on it is zero. The 3 forces acting on the rope are the two "pulls" F_1 and F_2 and the frictional force exerted by the post.

3.23 a) Choose bat as system. First condition gives $F - W = 0 \Rightarrow F = W = 8.91$ N

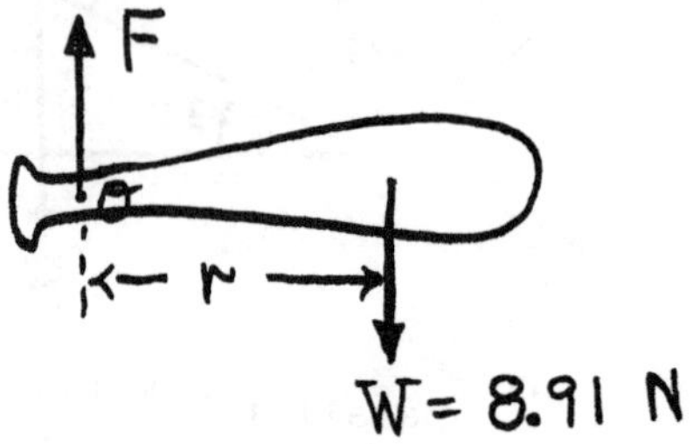

b) Choose O as torque axis.

Second condition gives

$$\sum \tau = F \cdot 0 - W \cdot r = -W \cdot r = -(8.91)(0.60) = -5.35 \text{ N} \cdot \text{m (minus} \Rightarrow \text{clockwise)}$$

This is torque on bat by gravity. For equilibrium, the player exerts an equal and opposite torque on the bat.

3.26 HINT: a) Choose the two balls collectively as a single body. Apply 1st condition to prove that $P_2 = 2W$ and that $P_1 = P_3$. Apply 2nd condition about torque axis through center of lower ball. b) Choose upper ball as body. Forces acting are P_1, W, and contact force. With W given and P_1 found from a), the 1st condition lets you determine the contact force.

3.28 HINT: The contact force between the barrel and the street (but not the curb) drops to zero when the barrel starts over the curb. Apply 2nd condition about the curb contact point. This choice of torque axis makes it unnecessary to determine the contact force exerted by the curb.

3.30 HINT: Use symmetry and apply 1st condition to entire roof to show $V = W/2$. Apply 2nd condition to left side of roof about apex point. Torque-producing forces are W/2 and the vertical and horizontal components of the

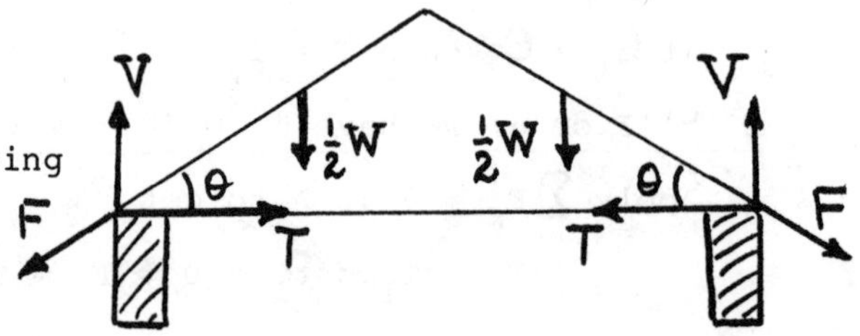

contact force exerted by the wall. The horizontal component equals the tension in the rod. Apply 2nd condition and result of a) to get $T = (W/4)\cot\theta$

3.31 HINT: Have fun with this one! Hi George.

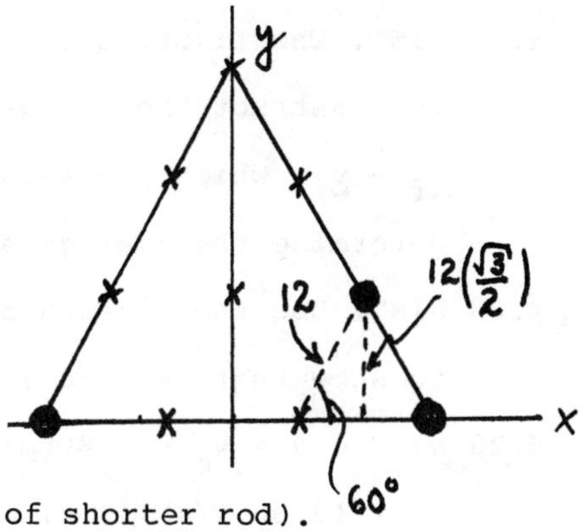

3.38 From the figure,

$\bar{x}$ = M(18+12-18)/3M = 4 in

$\bar{y}$ = M(0 + 12[$\sqrt{3}$/2] + 0)/3M = 2$\sqrt{3}$ = 3.46 in

3.43 HINT: Show that the maximum possible restoring torque (set up by the weight of the longer rod) is greater than the maximum tipping torque (set up by weight of shorter rod).

4.4 HINT: Use a vector diagram to convince yourself that $\mathbf{r}_A - \mathbf{r}_B$ is the position vector of B relative to A. Its magnitude gives the distance between A and B.

4.7 a) From the figure the displacement vector is $\Delta\mathbf{r} = 150\mathbf{i} + 65\mathbf{j}$ km

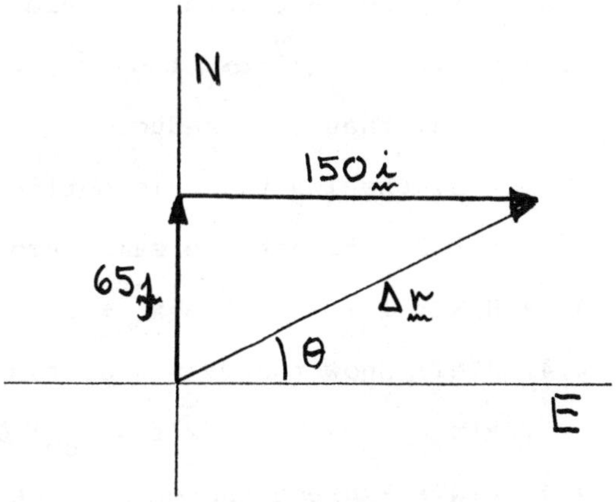

The average velocity is

$\bar{\mathbf{v}} = \Delta\mathbf{r}/\Delta t = (150\mathbf{i} + 65\mathbf{j})/2.5$ km/hr

$= 60\mathbf{i} + 26\mathbf{j}$ km/hr

In terms of magnitude and direction

$\bar{v} = [(60)^2 + (26)^2]^{\frac{1}{2}} = 65.4$ km/hr

$\theta = \arctan(65/150) = 23.4^\circ$ N of E

b) From a), $\bar{v} = 65.4$ km/hr

c) The average speed equals the total distance traveled (65+150 = 215 km) divided by the elapsed time (2.5 hr).

average speed = 215 km/2.5 hr = 86 km/hr

Note that the average speed differs from the magnitude of the average velocity.

4.18 $x = -4.9t^2$;

$\bar{v} = (x_f - x_i)/(t_f - t_i) = -4.9(t_f^2 - t_i^2)/(t_f - t_i)$

$\bar{v} = -4.9(t_f + t_i)$; With $t_i = 6$ s, $\bar{v} = -4.9(t_f + 6)$ m/s

a) $t_f = 6.1$ s, $\bar{v} = -59.29$ m/s; b) $t_f = 6.01$ s, $\bar{v} = -58.849$ m/s;

c) $t_f = 6.001$ s, $\bar{v} = -58.8049$ m/s; d) $t_f = 6.0001$ s, $\bar{v} = -58.80049$ m/s

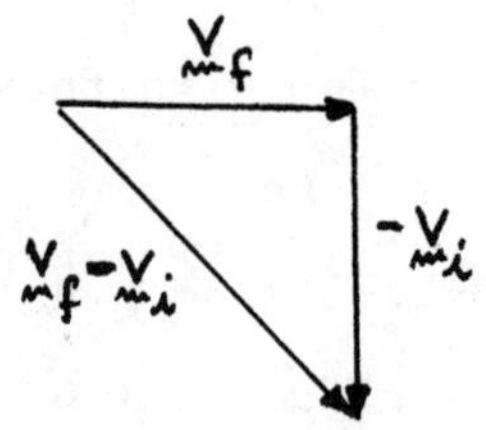

4.20 HINT: The figure shows you how to construct the change in velocity $\mathbf{v}_f - \mathbf{v}_i$, which you need to determine the average acceleration.

4.23 HINT: Use the algebraic manipulation employed in the solution of 4.18 to show that $a = 30 - 0.5(t_f + t_i)$ m/s^2

4.28 a) Use $v = v_o - 9.8t$ plus the fact that $v = 0$ when $t = 3$ s to find $v_o = (9.80 \text{ m/s}^2)(3 \text{ s}) = 29.4$ m/s

b) Use time symmetry: time of fall = time of rise = 3 s, and the fact that

distance up = distance down = $\frac{1}{2}gt^2 = \frac{1}{2}(9.80 \text{ m/s}^2)(3 \text{ s})^2 = 44.1$ m

4.31 HINT: a) Use $\bar{v}\Delta t = \Delta x$; b) With $v_o = 0$, $v = 2\bar{v}$; c) Use $x = \frac{1}{2}at^2$

4.34 HINT: For a constant acceleration, $\bar{v} = \frac{1}{2}(v_f + v_i)$

4.37 HINT: $x \propto gt^2 \Rightarrow t \sim \sqrt{x}$, which implies that if x is reduced by a factor nine, then t is reduced by a factor of $\sqrt{9} = 3$. If full-scale motion lasts 6 s, the 1/9 th scale motion lasts 2 s. At how many frames/s must you film for 2 s to get the same number of frames as in a 6-second full scale scene?

4.39 HINT: $1 + x + x^2 + x^3 + \ldots = \sum_{n=0}^{\infty} x^n = 1/(1-x)$ provided $|x| < 1$

4.41 HINT: Show that $v = 0$ at $t = t_{up} = -v_o/a$. Then show that $x(t = t_{up} - \Delta t) - x(t = t_{up} + \Delta t) = 0$ (for any Δt)

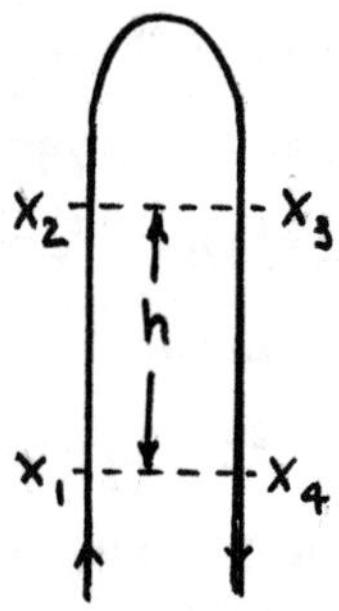

4.45 HINT: Express the quantities $x_2 - x_1 = h = x_3 - x_4$ in terms of v_o, g, and the 4 times. Eliminate the unknown v_o and solve for g in terms of h and the times.

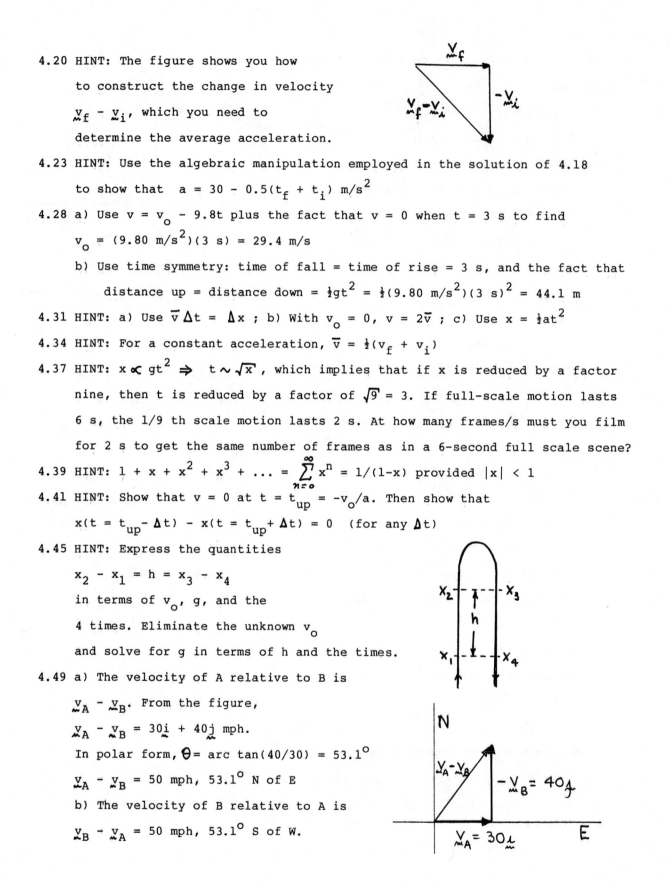

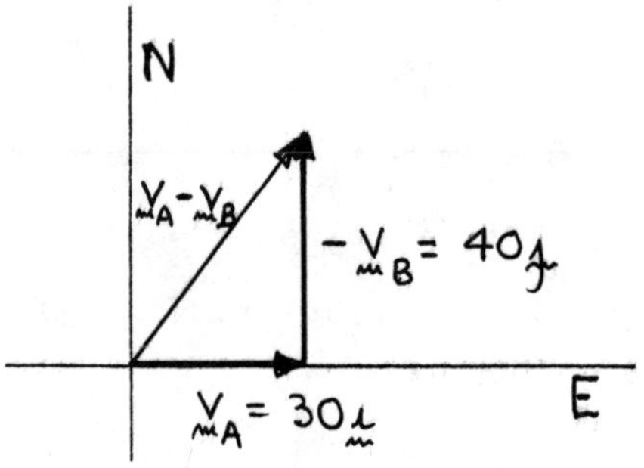

4.49 a) The velocity of A relative to B is $\mathbf{v}_A - \mathbf{v}_B$. From the figure,

$\mathbf{v}_A - \mathbf{v}_B = 30\mathbf{i} + 40\mathbf{j}$ mph.

In polar form, θ = arc tan(40/30) = 53.1°

$\mathbf{v}_A - \mathbf{v}_B$ = 50 mph, 53.1° N of E

b) The velocity of B relative to A is

$\mathbf{v}_B - \mathbf{v}_A$ = 50 mph, 53.1° S of W.

5.2 The vertical and horizontal components of motion are independent.

a) The initial velocity has no vertical component ($v_{yo} = 0$). Use Eq. 5.11 to find $t = \sqrt{2y/a_y} = \sqrt{2(1)/9.8} = 0.452$ s

b) The horizontal motion proceeds at a constant velocity of $v_{xo} = 2.3$ m/s. Use Eq. 5.7 with t = 0.452 s to get

$$x = v_{xo}t = (2.3 \text{ m/s})(0.452 \text{ s}) = 1.04 \text{ m}$$

5.7 HINT: y_{max} occurs when $v_y = 0$. Solve Eq. 5.9 for the value of t for which $v_y = 0$, and then substitute the result into Eq. 5.11 to get y_{max}. Use Eq. 5.13 to replace v_{yo} by $v_o \sin\theta$.

5.9 HINT: Use Eq. 5.17.

5.19 The time at which the altitude is a maximum is $t_m = -v_{yo}/a_y$. To prove that y is symmetric about the peak, we use Eq. 5.11 to prove that the difference $y(t_m+t_o) - y(t_m-t_o)$ is identically zero.

$$y(t_m+t_o) - y(t_m-t_o) =$$

$$v_{yo}[t_m+t_o-(t_m-t_o)] + \tfrac{1}{2}[(t_m+t_o)^2 - (t_m-t_o)^2]$$

$$= 2\,t_o[v_{yo} + a_y t_m] = 0\text{ , because } t_m = -v_{yo}/a_y$$

5.21 HINT: Use Eq 5.20. Consult Appendix 2 for the earth's radius.

6.5

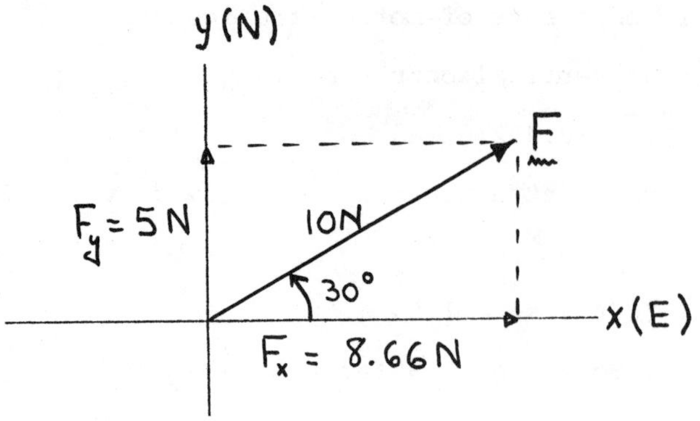

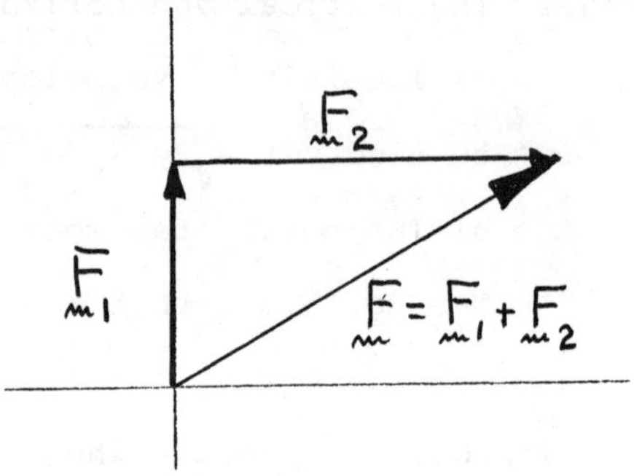

From $F_{net} = ma = 1\ kg \cdot 10\ m/s^2 = 10\ N$ we find that the magnitude of the net force is 10 N. Because $\underset{\sim}{F} = m\underset{\sim}{a}$ is a vector equation we conclude that $\underset{\sim}{F}$ has the same direction as $\underset{\sim}{a}$, that is 30° north of east. Thus, the net force is 10 N at 30° north of east. From the left figure we see that its components are $F_x = 10\cos 30^\circ = 8.66$ N (east) and $F_y = 10\sin 30^\circ = 5$ N (north).

The net force $\underset{\sim}{F}$ is the vector sum of $\underset{\sim}{F}_1$ (5N, north) and $\underset{\sim}{F}_2$, which we seek to determine. We have $\underset{\sim}{F}_1 + \underset{\sim}{F}_2 = \underset{\sim}{F}$, or $\underset{\sim}{F}_2 = \underset{\sim}{F} - \underset{\sim}{F}_1$. In component form,

$$F_{2x} = F_x - F_{1x} = 8.66 - 0 = 8.66\ N$$

$$F_{2y} = F_y - F_{1y} = 5 - 5 = 0\ N$$

We conclude that $\underset{\sim}{F}_2$ is directed along the +x-direction (east) and has a magnitude of 8.66 N, as suggested by the right figure above.

6.7 HINT: All possible vector sums of the 10 N force and a 15 N force can be constructed by generating a circle of "radius" 15 N with its center at the "tip" of the 10 N force vector.

6.10 The 15 N and 10 N forces combine to give a resultant force of 5 N south. The net force is the vector sum of a 20 N force east and a 5 N force south. The magnitude of the net force is $F = [(20)^2+(5)^2]^{\frac{1}{2}} = 20.6$ N

With m = 4 kg the acceleration is

$a = F/m = 20.6\ N/4\ kg = 5.15\ m/s^2$

The direction of $\underset{\sim}{a}$ is the same as that of $\underset{\sim}{F}$

$\theta = \arctan(5/20) = 14.0^\circ$ south of east

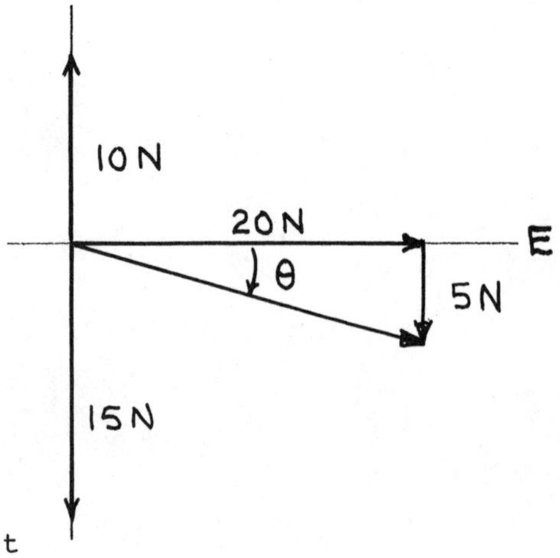

6.14 HINT: A constant force means that the acceleration is constant.

a) Use $a = \Delta v/\Delta t$; b) and c) Use $F = ma$

6.19 HINT: The distance from the center of the earth to a point at an altitude of 100 miles (= 161 km) is

$$r = 6.37 \times 10^6 \text{ m} + 0.161 \times 10^6 \text{ m} = 6.53 \times 10^6 \text{ m}$$

6.24 HINT: Use Kepler's third law ($T^2 = Kr^3$). You can avoid figuring out the value of K by taking ratios.

6.25 HINT: Kepler's third law gives the same value of T^2/r^3 for correct entries.

6.31 HINT: The force of gravity (weight) varies inversely as the square of the distance from the center of the earth. Moving from the surface to an altitude of 6370 km <u>doubles</u> that distance.

6.33 The ratio of weight to mass is $W/m = GM/r^2$. The centripetal acceleration is $v^2/r = (2\pi r/T)^2/r$. Equating and solving for r gives

$$r = [GM(T/2\pi)^2]^{1/3}$$
$$= [(6.67 \times 10^{-11})(5.98 \times 10^{24})(8.61 \times 10^4/2\pi)^2]^{1/3}$$
$$= 4.22 \times 10^7 \text{ m}$$

The altitude is

$$h = r - 6.37 \times 10^6 \text{ m} = 3.58 \times 10^7 \text{ m} = 22{,}200 \text{ miles}$$

6.38 $x = 6.2\ \text{ft}(12\ \text{in}/1\ \text{ft})(2.54\ \text{cm}/1\ \text{in})(1\ \text{m}/100\ \text{cm}) = 1.89\ \text{m}$

The spring constant is $k = F/x = mg/x = (0.1)(9.8)\ \text{N}/1.89\ \text{m} = 0.519\ \text{N/m}$

6.44 HINT: Use $ma = F_{net} = mg - f$ to express the frictional force in terms of m, a, and g. With f constant the acceleration is constant. The given data for displacement and time allow you to determine the acceleration.

6.49 HINT: Using $x = v^2/2a$, calculate the distance each car travels before coming to rest. Note also that the second car travels 3 m before its brakes are applied.

6.51 HINT: Use Eq. 6.17 to determine the terminal speed of the droplet.

6.58 a) Use the y-component of Newton's second law to determine N.

$$\sum F_y = N - mg\cos\theta = ma_y = 0$$

$$N = mg\cos\theta = 9800\cos 10^\circ$$

$$N = 9650\ N$$

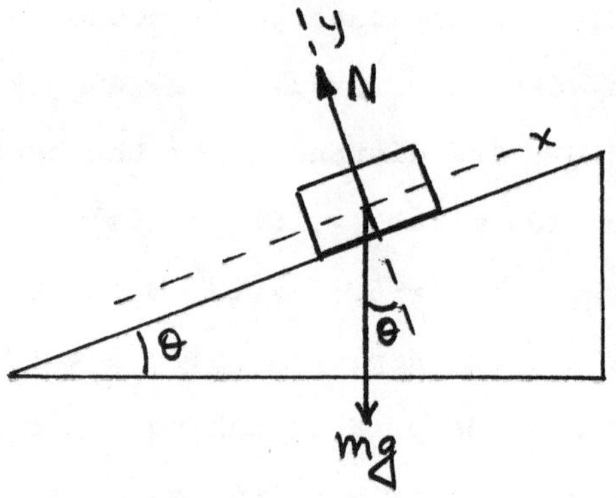

b) Use the x-component of Newton's second law to determine a_x.

$$\sum F_x = -mg\sin\theta = ma_x$$

$$a_x = -g\sin\theta = -1.70\ m/s^2$$ (minus indicates down the incline)

c) Use $v = v_o + a_x t$ with $v = 0$ to determine the stopping time.

$$0 = 25 - 1.7t \Rightarrow t = 25/1.7 = 14.7\ s$$

d) Use $v^2 = v_o^2 + 2a_x x$ with $v = 0$ to determine the stopping distance.

$$x = -(v_o^2/2a_x) = -625/2(-1.70) = 184\ m$$

6.62 HINT: Newton's second law takes the form

$$F_{net} = mg - N = ma$$

where N is the spring scale reading, and down is taken to be positive.

6.66 If R is expressed in AU and T is expressed in years, then Kepler's third law takes the form $T^2 = R^3$. You can verify this for the earth: T = 1 yr, R = 1 AU. For Pluto, $T = R^{3/2} = (39.44)^{3/2}$ yr = 248 yr

7.3 a) $W_N = \mathbf{N}\cdot\Delta\mathbf{s} = 0$ because $\mathbf{N}$ and $\Delta\mathbf{s}$ are perpendicular.

b) $W_g = m\mathbf{g}\cdot\Delta\mathbf{s} = -mg\cdot\sin\theta\cdot\Delta s$

$$= -(1000)(9.80)(\sin 3^\circ)(100)$$

$$= -5.13 \times 10^4\ J$$

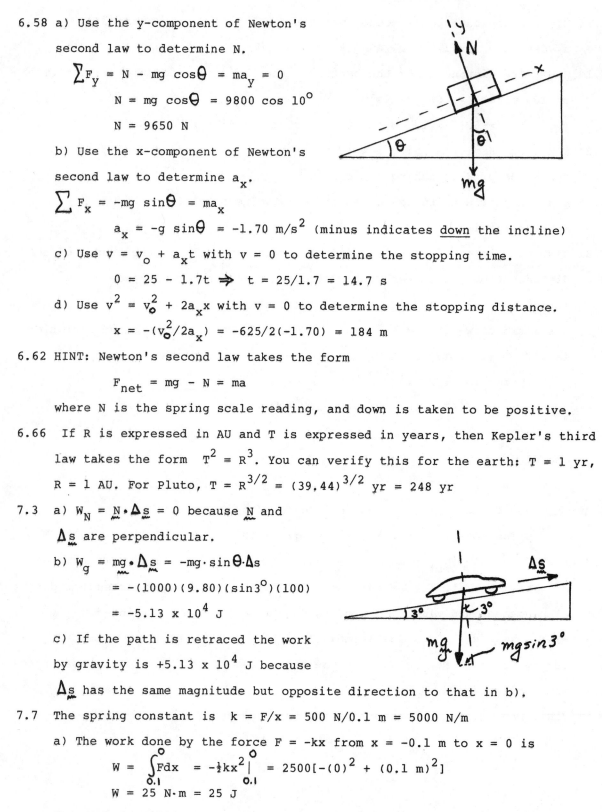

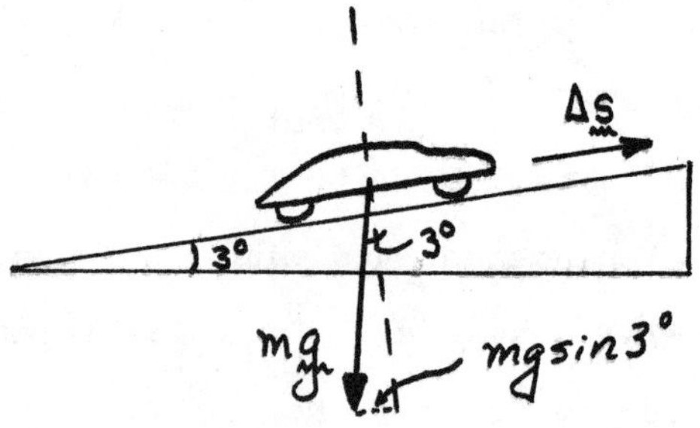

c) If the path is retraced the work by gravity is $+5.13 \times 10^4$ J because $\Delta\mathbf{s}$ has the same magnitude but opposite direction to that in b).

7.7 The spring constant is $k = F/x = 500\ N/0.1\ m = 5000\ N/m$

a) The work done by the force $F = -kx$ from $x = -0.1$ m to $x = 0$ is

$$W = \int_{0.1}^{0} F\,dx = -\tfrac{1}{2}kx^2\Big|_{0.1}^{0} = 2500[-(0)^2 + (0.1\ m)^2]$$

$$W = 25\ N\cdot m = 25\ J$$

The work is positive because the force and displacement are parallel.

7.7 b) The work has the same magnitude as in a) but is negative because the force and displacement are in opposite directions. Formally,

$$W = \int_0^{0.1} -kx\,dx = -\tfrac{1}{2}kx^2 \Big|_0^{0.1} = -25 \text{ J}$$

c) Adding the works of a) and b) shows that the total work done by the spring force from x = -0.1 m to +0.1 m is zero.

7.8 HINT: $W = \int_{x_o}^{0} F\,dx$ is positive because F and dx are in the same direction throughout the motion from x_o to 0.

7.14 b) HINT: $\int_{r_1}^{r_2} dr/r^2 = (1/r_1) - (1/r_2)$

7.15 HINT: Let $r/\sigma = x$ and note that the work integral can be evaluated as

$$\int_{r_1}^{r_2} F\,dr = \sigma F_o \int_{r_1/\sigma}^{r_2/\sigma} (2x^{-13} - x^{-7})\,dx = (\sigma F_o/6)(x^{-6} - x^{-12}) \Big|_{r_1/\sigma}^{r_2/\sigma}$$

7.18 HINT: The work done by gravity depends on the vertical displacement.

7.22 a) With k = 3 N/m and a = 1 m, $\underset{\sim}{F}\cdot d\underset{\sim}{s} = -k(y\,dx + a\,dy) = -3(y\,dx + dy)$ J

Along the vertical leg from (0,0) to (0,2) x is constant so dx = 0 and $\underset{\sim}{F}\cdot d\underset{\sim}{s} = -3dy$ J

The work done is $-3\int_0^2 dy = -6$ J

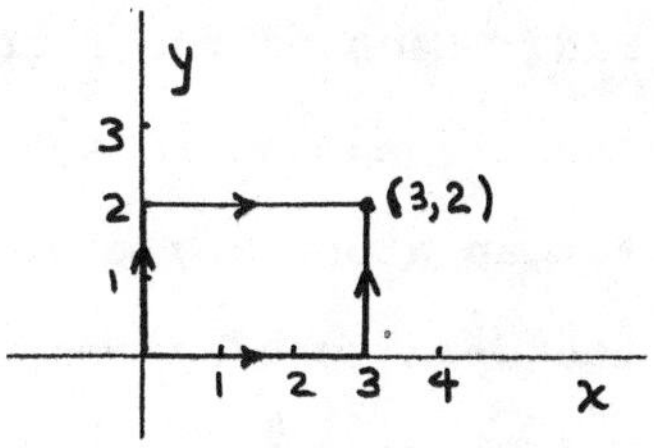

Along the horizontal leg from (0,2) to (3,2) y = 2 m is constant so dy = 0 and $\underset{\sim}{F}\cdot d\underset{\sim}{s} = -3y\,dx = -6dx$ J. The work done is

$$-6\int_0^3 dx = -18 \text{ J}$$

The total work is the sum of the works along the two legs: W = -24 J

b) You should be able to show W = -6 J. The fact that this result differs from that for the path of part a) shows that the force is non-conservative. (See Section 8.1)

7.29 HINT: $v = dx/dt$; $K = \tfrac{1}{2}mv^2$

7.32 The work done by the spring is $\tfrac{1}{2}kx^2$. The kinetic energy increases from zero to $\tfrac{1}{2}mv^2$. The work-energy principle takes the form

$$\tfrac{1}{2}kx^2 = \tfrac{1}{2}mv^2$$

which gives

$$v = [kx^2/m]^{\frac{1}{2}} = [100(.1)^2/.05]^{\frac{1}{2}} = 4.47 \text{ m/s}$$

7.34 HINT: $\Delta K = 0$ because the speed remains constant. It follows from the work-energy principle that the net work done is zero.

7.38 HINT: b) $\Delta K = 0$ so the sum of the work by gravity (+) and by the spring (-) is zero.

c) Use the work-energy principle to show

$$v^2 = 2g(y+x) - kx^2/m$$

The maximum value of v follows from $d(v^2)/dx = 0$.

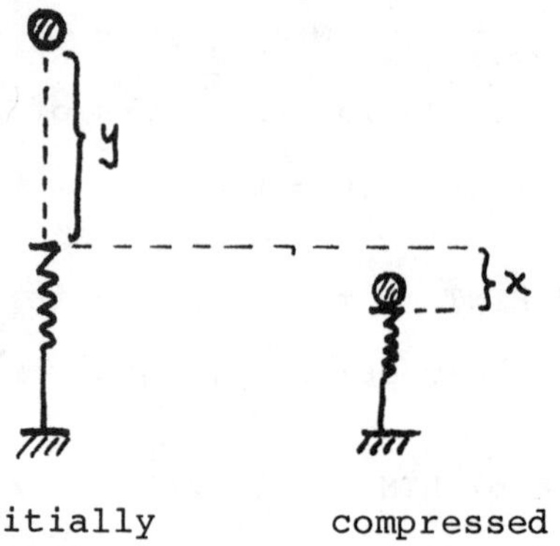

7.43 $P = W/t = \Delta K/t = \frac{1}{2}mv^2/t$; $v = 60$ mph(0.447 m/s/mph) = 26.8 m/s

$P = \frac{1}{2}(1000)(26.8)^2$ J/10 s $= 3.59 \times 10^4$ W (1 hp/746 W) = 48.1 hp

7.50 HINT: The force the man exerts equals the tension in the rope. With 2 strands supporting his weight the tension equals half his weight.

7.53 HINT: Each of the four supporting rope segments is shortened by 1 m when the "pull" segment is shortened by 4 m.

8.1 For an arbitrary closed path (ABCDA) the work done by a conservative force $\mathbf{F}$ is zero. $\oint \mathbf{F}\cdot d\mathbf{s} = 0$

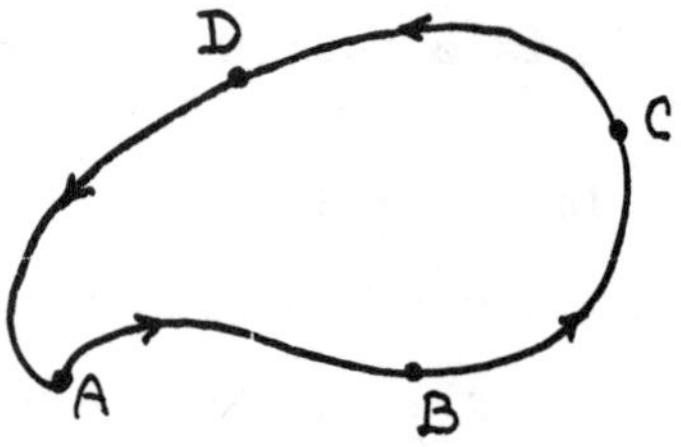

Viewing the closed path as the two legs ABC and CDA we have $\int_{ABC} \mathbf{F}\cdot d\mathbf{s} + \int_{CDA} \mathbf{F}\cdot d\mathbf{s} = 0$, or $\int_{ABC} \mathbf{F}\cdot d\mathbf{s} = -\int_{CDA} \mathbf{F}\cdot d\mathbf{s}$

Reversing the direction along which the path is traced changes the sign of the integral: $\int_{ADC} \mathbf{F}\cdot d\mathbf{s} = -\int_{CDA} \mathbf{F}\cdot d\mathbf{s}$ [This is the 3-dim version of $\int_a^b F dx = -\int_b^a F dx$]

With the path reversal and change of sign we have

$$\int_{ABC} \mathbf{F}\cdot d\mathbf{s} = \int_{ADC} \mathbf{F}\cdot d\mathbf{s}$$

This states that the integral $\int \mathbf{F}\cdot d\mathbf{s}$ has the same value for the paths ABC and ADC. But ABC and ADC are arbitrary paths from A to C, so the integral has the same value for all paths from A to C - it is path independent.

8.5 HINT: $W = F\Delta x$

8.6 $W = \int_0^x F dx = \int_0^x (100x - 100)dx = 50x^2 - 100x = 50x(x-2)$

a) W is zero when x = 0 (no motion) and when x = 2 m.

b) Although the given force is conservative, the fact that W = 0 when x = 2 m does not prove that the force is conservative. To qualify as conservative the work done must be zero over ALL CLOSED PATHS.

8.8 $U = \frac{1}{2}kx^2$; $\Delta U = \frac{1}{2}k[x_f^2 - x_i^2] = \frac{1}{2}(1000)[(0.2)^2 - (0.05)^2] = 18.8$ J

8.9 HINT: $U = \frac{1}{2}kx_1^2 + \frac{1}{2}kx_2^2 + \frac{1}{2}kx_3^2$; $x_2 = x_3 = R\Delta\theta = R\pi/6$

8.17 HINT: $g = GM_E/R_E^2$

8.21 HINT: Let $(\sigma/r)^6 = x$. This substitutuion converts the equation for U into a quadratic equation in x. You should find the roots to be x = 0.233 and x = 0.767.

8.24 $K_i + U_i = K_f + U_f$

$0 + 0 = 0 + \frac{1}{2}kx^2 - mgx$; $x = 2mg/k = 2(15)(9.80)/1000 = 0.294$ m

8.27 HINT: Both masses move at the same speed. The total energy can be conveniently set equal to zero. ($K_i = U_i = 0$)

8.31 HINT: At the top the net force is radial and equals the tension + mg. At the bottom the net force is radial and equals the tension - mg.

8.35 $mv^2/r = GM_Em/r^2$; $K = \frac{1}{2}mv^2 = \frac{1}{2}GM_E/r$

$E = K + U = \frac{1}{2}GM_E/r - GM_E/r = -\frac{1}{2}GM_E/r$

8.37 HINT: Ignore the initial gravitational potential energy, that is, the gravitational potential energy due to the earth.

8.55 $F = -dU/dx = -d(x^3)/dx = -3x^2 = 0$ at x = 0. If the particle is given a small displacement to the left (x < 0) it continues to the left. If the particle is given a small displacement to the right (x > 0) it slows down, comes to rest momentarily, then moves back toward x = 0. It continues to the left past the x = 0 equilibrium point. Because all displacements from x = 0 lead to continued motion to the left, the equilibrium at x = 0 is unstable.

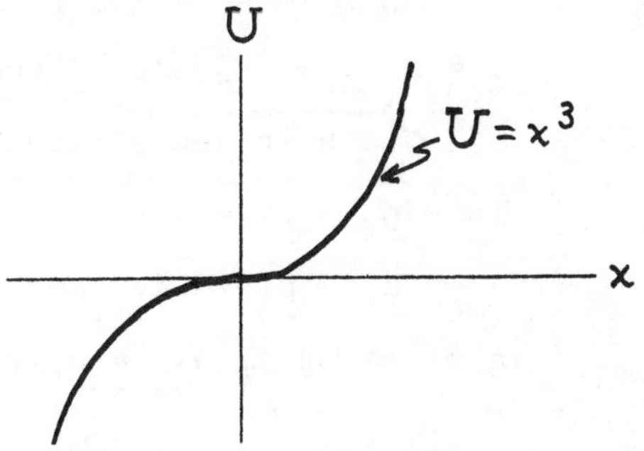

8.58 HINT: $U = 5x^2$; $K = E - U = 20 - 5x^2$ J

9.2 $F\Delta t = mg\,\Delta t = (200)(9.80)(.80) = 1570\ \text{N}\cdot\text{s}$

9.5 J = "Area" = $\frac{1}{2}(1200\ \text{N})(0.1\ \text{s}) = 60\ \text{N}\cdot\text{s}$

9.11 HINT: b) For the geometry shown →

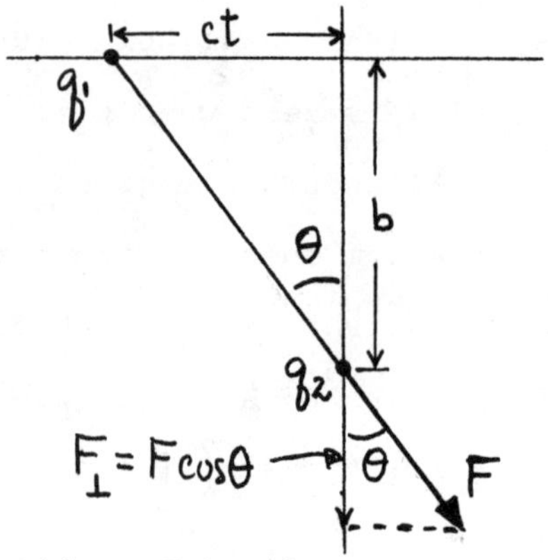

$$J_\perp = \int_{-\infty}^{+\infty} F_\perp\,dt = kq_1q_2 \int_{-\infty}^{+\infty} \cos\theta\,dt/r^2$$

You can express r and $\cos\theta$ in terms of the integration variable t. Let $t = 0$ correspond to the moment when $r = b$.

9.13 HINT: Use $v = v_o + at$ and carry out the integration. Note that

$v_i = v_o + at_i$; $v_f = v_o + at_f$

9.15 $F = dp/dt = d(mA\sin\omega t)/dt = -m\omega A\sin\omega t$

a) At $\omega t = 90^\circ$, $\sin\omega t = 1$, $F = -m\omega A$

b) At $\omega t = 0^\circ$, $\sin\omega t = 0$, $F = 0$

9.19 HINT: $\Delta p = mv - 0$; $J = \int_0^t F dt = F\int_0^t dt = Ft$

9.21 The objects exert equal and opposite forces on each other and so receive equal but opposite linear impulses.

$$m_1\Delta v_1 = J = -m_2\Delta v_2 \Rightarrow -(\Delta v_1/\Delta v_2) = m_2/m_1,$$

which shows that the ratio of the masses determines the ratio of velocity changes.

9.23 HINT: Note that the linear impulse is made up of a positive contribution during the first 5 s and a negative contribution during the last 1 s.

9.25 $p_i = mv$ (scalar)

$\underset{\sim}{p}_i = -mv\underset{\sim}{i}$ (vector)

$p_f = mv$ (scalar)

$\underset{\sim}{p}_f = mv[-\underset{\sim}{i}\tfrac{1}{2} + \underset{\sim}{j}\sqrt{3}/2]$ (vector)

↖ a unit vector at 120° from +x-axis

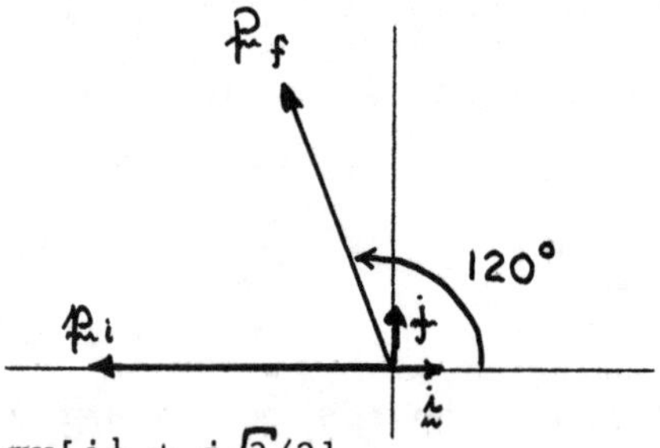

$\Delta\underset{\sim}{p} = \underset{\sim}{p}_f - \underset{\sim}{p}_i = mv[-\underset{\sim}{i}\tfrac{1}{2} + \underset{\sim}{j}\sqrt{3}/2] - (-mv\underset{\sim}{i}) = mv[\underset{\sim}{i}\tfrac{1}{2} + \underset{\sim}{j}\sqrt{3}/2]$

$J = |\underset{\sim}{p}_f - \underset{\sim}{p}_i| = mv$ because $|[\underset{\sim}{i}\tfrac{1}{2} + \underset{\sim}{j}\sqrt{3}/2]| = 1$

In terms of J: $\Delta\underset{\sim}{p} = \underset{\sim}{p}_f - \underset{\sim}{p}_i = J[\underset{\sim}{i}\tfrac{1}{2} + \underset{\sim}{j}\sqrt{3}/2]$

9.28 $p_i = 0 = p_f = m_{Th}v_{Th} - m_{alpha}v_{alpha}$

$v_{Th} = m_{alpha}v_{alpha}/m_{Th} = (4)(1.50 \times 10^7 \text{ m/s})/234 = 2.56 \times 10^5 \text{ m/s}$

9.31 HINT: Show that $\underset{\sim}{p}_f = 1050\underset{\sim}{i} + 900\underset{\sim}{j}$ kg•m/s and use conservation of linear momentum.

9.34 a) $p_i = Mv + 2M\cdot 0 = Mv$; $p_f = 3Mv_f$

Conservation of linear momentum, $p_i = p_f$, gives $v_f = v/3$.

The change in speed is $\Delta v = v - v_f = 2v/3$

b) Here the change in speed is v/3.

9.38 $p_i = 0 = p_f = p_1 + p_2 \Rightarrow p_1 = -p_2$

The kinetic energies can be expressed in terms of the momenta

$K_1 = p_1^2/2m_1$; $K_2 = p_2^2/2m_2$. Forming the ratio K_1/K_2 we can cancel the ratio of momentum squares because $p_1 = -p_2$,

$K_1/K_2 = (p_1^2/2m_1)/(p_2^2/2m_2) = m_2/m_1$

10.2 $(M_{Moon} + M_{Earth})x_{cm} = M_{moon}x_{moon} + M_{Earth}x_{Earth}$

With origin at the center of the earth, $x_{Earth} = 0$. Solving for x_{cm},

$x_{cm} = x_{Moon}/(1 + M_{Earth}/M_{Moon}) = 3.84 \times 10^8 \text{ m}/[1 + 1/0.0123]$

$x_{cm} = 4.67 \times 10^6$ m This is about 2/3 the radius of the earth.

10.6 $y_{cm} = 0$, $z_{cm} = 0$ because $y_1 = y_2 = z_1 = z_2 = 0$. Let x_2 denote the larger of x_1 and x_2. Then

$$x_{cm} = (m_1x_1 + m_2x_2)/(m_1 + m_2)$$
$$= x_1 + (m_2/m_1+m_2)(x_2-x_1) > x_1$$

and

$$x_{cm} = x_2 + (-m_1/m_1+m_2)(x_2-x_1) < x_2$$

which shows that the center of mass lies between x_1 and x_2, that is, between the two particles.

10.12 a) $x_{cm} = \int x\,dm / m$

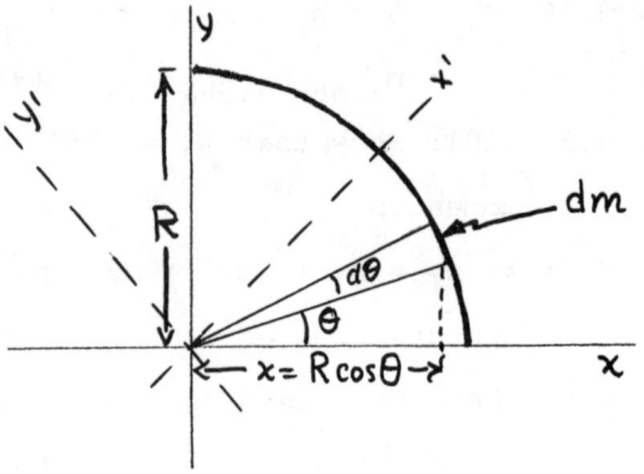

$dm = (m/L)ds = (m/\pi R/2)\, R d\theta$

$x_{cm} = (2/\pi) \int_0^{\pi/2} x\,d\theta$; $x = R\cos\theta$

$x_{cm} = (2R/\pi) \int_0^{\pi/2} \cos\theta\, d\theta = (2R/\pi) \sin\theta \Big|_0^{\pi/2}$

$x_{cm} = 2R/\pi = y_{cm}$, by symmetry.

b) For a set of symmetry axes (x',y') you can see that $y_{cm} = 0$, by symmetry. The integral for x_{cm} is set up as above in a) but the limits run from $-\pi/4$ to $+\pi/4$, with the result $x_{cm} = 2\sqrt{2}\,R/\pi$

10.17 The center of mass of the two-body system does not move. The capsule moves a distance x = 0.02 m while the astronaut moves a distance 20 - x in the opposite direction. The fact the center of mass does not move is expressed as

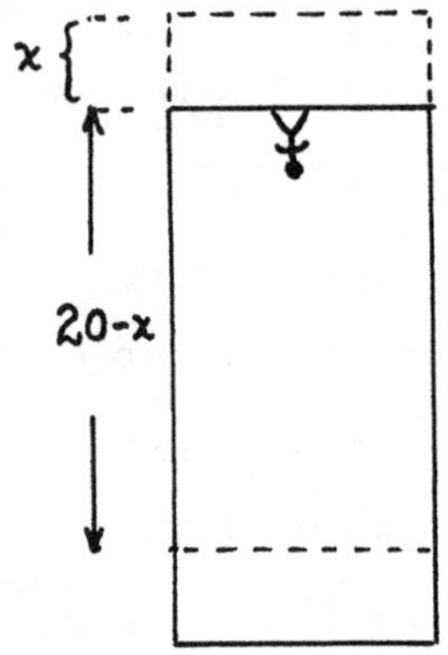

$\Delta x_{cm} = Mx - 80(20 - x) = 0$

Solving for M gives

$$M = 80 \text{ kg } (20 - x)/x = 80 \text{ kg } (20 - 0.02)/0.02 = 79{,}900 \text{ kg}$$

10.22 Problem 10.21 shows that the 228 nucleus recoils at 0.0232×10^7 m/s, when the decaying 232 nucleus is at rest. In order for the the 228 nucleus to end up at rest, the about-to-decay 232 nucleus would have to be moving at 0.0232×10^7 m/s. The alpha would then move at $(1.32 + 0.0232) \times 10^7$ m/s $= 1.34 \times 10^7$ m/s relative to the lab and residual 228 nucleus.

10.27 a) Lab Frame: $K_{1i} = \frac{1}{2}8(6)^2 = 144$ J; $K_{2i} = \frac{1}{2}2(-4)^2 = 16$ J

$K_i = K_{1i} + K_{2i} = 160$ J

$K_{1f} = \frac{1}{2}8(2)^2 = 16$ J; $K_{2f} = \frac{1}{2}2(12)^2 = 144$ J;

$K_f = K_{1f} + K_{2f} = 160$ J. This verifies $K_i = K_f$ and thus that $\Delta K = K_f - K_i = 0$ in the lab frame.

b) Center of Momentum Frame: $K_{1i} = \frac{1}{2}8(2)^2 = 16$ J;

$K_{2i} = \frac{1}{2}2(-8)^2 = 64$ J; $K_i = 80$ J;

$K_{1f} = \frac{1}{2}8(-2)^2 = 16$ J; $K_{2f} = \frac{1}{2}2(8)^2 = 64$ J; $K_f = 80$ J.

This verifies that $K_i = K_f = 80$ J and that $\Delta K = K_f - K_i = 0$ in the center-of-momentum frame. Even though the total kinetic energy is different in the two frames, the conservation of energy principle ($\Delta K = 0$) is valid in both frames. The velocities in the lab and center-of-momentum frames are related by a Galilean transformation. We say that the principle of energy conservation is <u>invariant</u> under (unchanged by) a Galilean transformation.

10.36 From Eq. 10.44 $\tan\theta = \sin\theta'/(\cos\theta' + m_1/m_2)$

With $m_1 = m_2$ this becomes $\tan\theta = \sin\theta'/\cos\theta' + 1$

Next we need the identities: $\cos\theta' = \cos^2(\theta'/2) - \sin^2(\theta'/2)$

$1 = \cos^2(\theta'/2) + \sin^2(\theta'/2)$

Adding shows $\cos\theta' + 1 = 2\cos^2(\theta'/2)$

The other identity we need is: $\sin\theta' = 2\sin(\theta'/2)\cos(\theta'/2)$

Inserting these expressions for $\sin\theta'$ and $\cos\theta' + 1$ into the equation for $\tan\theta$ gives

$$\tan\theta = [2\sin(\theta'/2)\cos(\theta'/2)]/[2\cos^2(\theta'/2)]$$
$$= \sin(\theta'/2)/\cos(\theta'/2) = \tan(\theta'/2)$$

which shows the desired result, $\theta = \theta'/2$

10.38 HINT: Let u denote the speed of the car at the moment the boy and girl jump. If the jumpers move at speed u_o relative to the car, they move at speed $u - u_o$ relative to the earth. In applying momentum conservation you need an inertial frame of reference such as the earth. The car is accelerated and is not an inertial frame.

10.41 Using $v = v_o + u_o \ln(m_o/m)$, with $v_o = 0$ and $v = u_o$ gives

$\ln(m_o/m) = 1$; $m_o/m = e = 2.718...$; $m/m_o = 0.368$

The fraction of the mass ejected is $1 - (m/m_o) = 0.632$

11.2 a) Because $\mathbf{r}$ and $\mathbf{p}$ are parallel, the angular momentum $\mathbf{r} \times \mathbf{p}$ is zero.

b) $L = r_{\perp} mv = (100\ \text{m})(1500\ \text{kg})(80000\ \text{m/hr})(1\ \text{hr}/3600\ \text{s})$

$= 3.33 \times 10^6\ \text{kg}\cdot\text{m}^2/\text{s}$

11.6 HINT: Use Eq. 11.10.

11.10 We can prove that the angular momentum of the alpha particle is a constant by showing that its time derivative $d\mathbf{L}/dt$ is zero.

$d\mathbf{L}/dt = d(\mathbf{r} \times \mathbf{p})/dt = (d\mathbf{r}/dt) \times \mathbf{p} + \mathbf{r} \times (d\mathbf{p}/dt)$

$d\mathbf{r}/dt = \mathbf{v}$, the particle velocity;

$d\mathbf{p}/dt = \mathbf{F}$, the force exerted on the alpha particle by the nucleus.

Substituting into $d\mathbf{L}/dt$ gives $d\mathbf{L}/dt = \mathbf{v} \times \mathbf{p} + \mathbf{r} \times \mathbf{F}$

The term $\mathbf{v} \times \mathbf{p}$ is zero because $\mathbf{v}$ and $\mathbf{p} = m\mathbf{v}$ are parallel and so their cross product is zero. The term $\mathbf{r} \times \mathbf{F}$ is also zero, because the electric force exerted on the alpha particle by the nucleus is always directed parallel to the position vector $\mathbf{r}$. In other words, $\mathbf{r}$ and $\mathbf{F}$ are always parallel and so their cross product is zero. With $\mathbf{v} \times \mathbf{p} = 0$ and $\mathbf{r} \times \mathbf{F} = 0$ we have the desired result: $d\mathbf{L}/dt = 0$. The fact that $d\mathbf{L}/dt = 0$ means that $\mathbf{L}$ is constant. As a constant **vector** both the magnitude and direction of $\mathbf{L}$ must remain constant.

11.11 From the figure, the area swept out in time Δt is half the area of the parallelogram with sides $\mathbf{r}$ and $\mathbf{r} + \Delta\mathbf{r}$. This is half of the magnitude of the cross product of $\mathbf{r}$ and $\mathbf{r} + \Delta\mathbf{r}$.

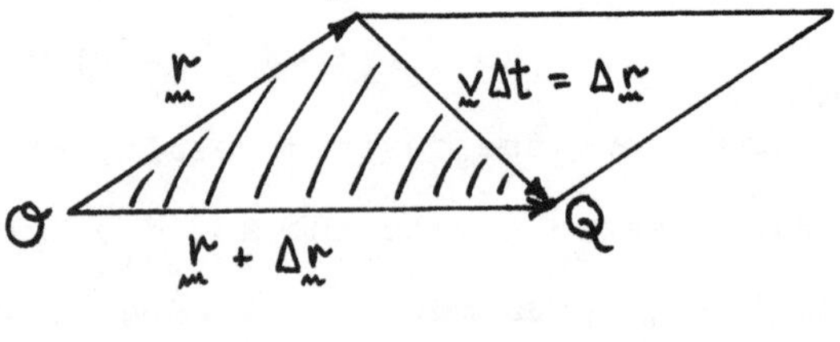

$$\Delta A = \tfrac{1}{2}|\mathbf{r} \times (\mathbf{r} + \Delta\mathbf{r})|$$

With $\Delta\mathbf{r} = \mathbf{v}\Delta t$, $A = \tfrac{1}{2}|\mathbf{r} \times \mathbf{r} + \mathbf{r} \times \mathbf{v}\Delta t| = \tfrac{1}{2}|\mathbf{r} \times \mathbf{v}|\Delta t$

The rate at which area is swept out is $\Delta A/\Delta t = \tfrac{1}{2}|\mathbf{r} \times \mathbf{v}|$

11.13 HINT: The bob is in uniform circular motion in a circle of radius $\ell\sin\theta$. Use Newton's second law to relate the radial acceleration $v^2/\ell\sin\theta$ to the radial component of the tension in the cord.

11.18 $L_e = M_e r_e v_e$; $L_m = M_m r_m v_m$

The speed of the center of the earth relative to the center of mass equals the speed of the center of mass relative to the center of the earth.

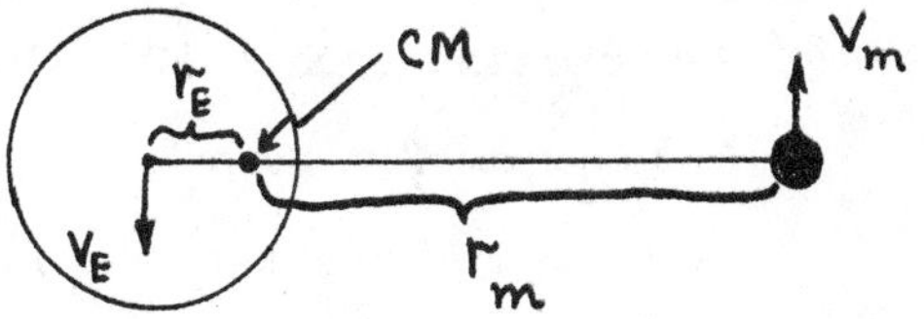

$v_e = 2\pi r_e/T$, where $T = 2.36 \times 10^6$ s is the orbital period of the earth-moon system about the center of mass. The speed of the moon is $v_m = 2\pi r_m/T$ so that $L_e = [M_e r^2]2\pi/T$; $L_m = [M_m r^2](2\pi/T)$

The data are: $M_e = 5.98 \times 10^{24}$ kg; $r_e = 4.66 \times 10^6$ m

$M_m = 7.35 \times 10^{22}$ kg; $r_m = 3.79 \times 10^8$ m

These give $L_e = 3.46 \times 10^{32}$ kg m^2/s; $L_m = 2.81 \times 10^{34}$ kg m^2/s

11.21 HINT: The total angular momentum is twice that of one particle ($L = 2mvr$).

11.26 HINT: a) See Example 1. b) The result given on page 224,

$$L_m^2 = (GM_m^2 M_E)r$$

is helpful.

11.29 HINT: Remember that $M_E r_E = M_M r_M$ for distances measured relative to the center of mass.

11.31 a) The statement of the problem assumes that the two crates move in circular orbits with the same period. Newton's second law for each crate can be expressed as [1] $ma_1 = GM_E m/r_1^2 - T$

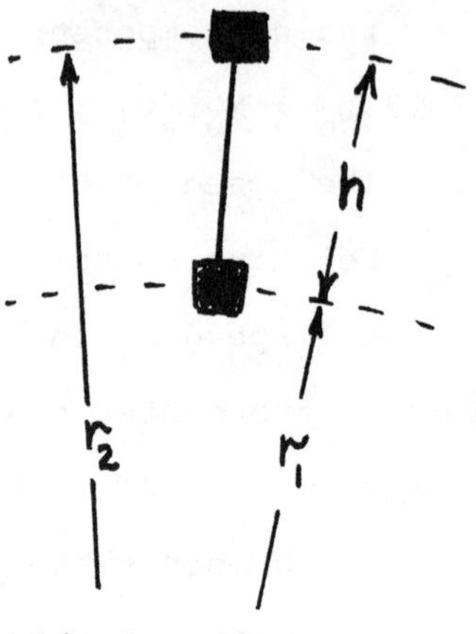

[2] $ma_2 = GM_E m/r_2^2 + T$

Subtracting equals [2] - [1] gives

$2T + GM_E m/r_2^2 - GM_E m/r_1^2 = m(a_2 - a_1)$

Solving for the tension T gives

$T = \frac{1}{2}m(a_2 - a_1) + \frac{1}{2}GM_E m/r_1^2\{1 - [1 + h/r_1]^{-2}\}$

On the right side we have made use of $r_2 = r_1 + h$

The assumption of equal periods lets us set

$a_1 = r_1(2\pi/P)^2$; $a_2 = r_2(2\pi/P)^2$, where P is the orbital period

This step gives $a_2 - a_1 = [r_2 - r_1](2\pi/P)^2 = h(2\pi/P)^2$

Because the tension is expected to be small compared to the force of gravity exerted by the earth we have approximately $mr_1(2\pi/P)^2 = GM_E m/r_1^2$ so that $(2\pi/P)^2 = GM_E/r_1^3$ which converts the right side of the equation for T into

$$T = \tfrac{1}{2}GM_E mh/r_1^3 + GM_E m/r_1^2\{\cancel{1} - [\cancel{1} - 2h/r_1]\}$$

The term in [ ] is the result of expanding $[1 + h/r_1]^{-2}$ using the binomial expansion, and retaining only the first two terms. The final result is

$$T = (3/2)GM_E mh/r_1^3$$

b) For h = 20 m, m = 1 kg, $r_1 = 6.37 \times 10^6$ m + 200 km = 6.57×10^6 m

$$T = 4.22 \times 10^{-5} \text{ N}$$

c) The gravitational force between the two crates is

$$F = Gm^2/h^2 = (6.67 \times 10^{-11})(1)^2/(20)^2 = 1.67 \times 10^{-13} \text{ N}$$

d) The tension and mutual gravitational forces are equal when

$$Gm^2/h^2 = (3/2)GM_E m/r_1^3$$

Solving for r_1 gives

$$r_1 = h[3M_E/2m]^{1/3} = 20[3(5.98 \times 10^{24})/2(1)]^{1/3}$$

$$= 4.16 \times 10^9 \text{ m}$$

12.5 The hour hand makes one revolution in 12 hours. Its angular velocity is

$$\omega = \Delta\theta/\Delta t = 2\pi\ \text{rad}/[12\ \not{hr}(3600\ \text{s}/\not{hr})] = 1.45 \times 10^{-4}\ \text{rad/s}$$

The minute hand makes one revolution in 1 hour. Its angular velocity is

$$\omega = \Delta\theta/\Delta t = 2\pi\ \text{rad}/[1\ \not{hr}(3600\ \text{s}/\not{hr})] = 1.75 \times 10^{-3}\ \text{rad/s}$$

The second hand makes one revolution in 1 minute. Its angular velocity is

$$\omega = \Delta\theta/\Delta t = 2\pi\ \text{rad}/[1\ \not{min}(60\ \text{s}/\not{min})] = 0.105\ \text{rad/s}$$

12.6 HINT: In order to present the same hemisphere to an observer at the center of its orbit, the orbital period of the moon must equal its rotational (spin) period. [Tidal friction produces and maintains this equality.]

12.10 a) 1 sidereal year = 3.16×10^7 s. To convert from 1 rev/yr to rad/s we note that 1 revolution corresponds to 2π radians.

$$\omega = 1\ \not{rev}/\not{yr}(2\pi\ \text{rad}/\not{rev})(1\ \not{yr}/3.16 \times 10^7\ \text{s}) = 1.99 \times 10^{-7}\ \text{rad/s}$$

b) $v = r\omega = (1.50 \times 10^{11}\ \text{m})(1.99 \times 10^{-7}\ \text{rad/s}) = 29.9\ \text{km/s}$

12.12 We can use the equation which defines angular acceleration in terms of the change in angular velocity.

$$\alpha = \Delta\omega/\Delta t = (\omega_f - \omega_i)/\Delta t$$

The final angular velocity is zero. The initial angular velocity is related to the speed of the car (v) and the radius of the track (r) by

$$\omega = v/r = 2\pi v/2\pi r = 2\pi v/C \qquad \text{where } C = 2\pi r \text{ is the circumference}$$

$$= [2\pi\ \text{rad} \cdot 175\ \text{mph} \cdot (0.447\ \not{m}/\text{s/mph})]/2 \times 10^3\ \not{m} = 0.246\ \text{rad/s}$$

The change in angular velocity from 0.246 rad/s to zero takes place in a time of $\Delta t = 20$ s so

$$\alpha = (0 - 0.246\ \text{rad/s})/20\ \text{s} = -0.0123\ \text{rad/s}^2$$

12.19 HINT: The ratio of spin energy to translational energy is independent of the mass. The ball spins about its center of mass ($I_{cm} = 2mr^2/5$).

12.22 $K_1 = \frac{1}{2}mv_1^2$; $K_2 = \frac{1}{2}mv_2^2$

The masses rotate with the same period and thus the same angular velocity. Thus $v_1 = r_1\omega$ and $v_2 = r_2\omega$, which gives

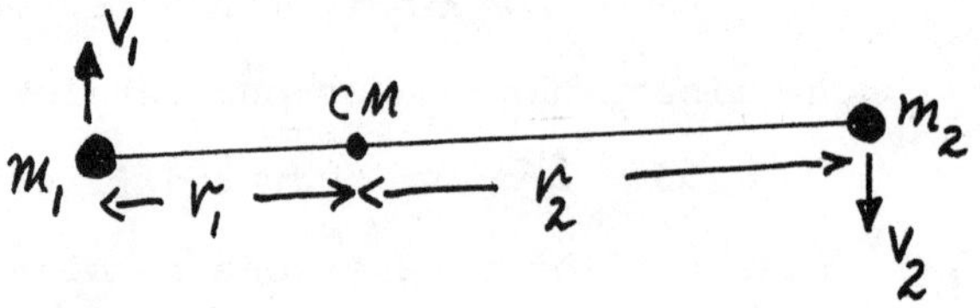

$K_1/K_2 = \frac{1}{2}m_1(r_1\omega)^2/\frac{1}{2}m_2(r_2\omega)^2 = m_1r_1^2/m_2r_2^2$

The distances r_1 and r_2 are related by $m_1r_1 = m_2r_2$; $(r_1/r_2)^2 = (m_2/m_1)^2$. This gives $K_1/K_2 = m_2/m_1$. From Ex 9, the masses are $M_E = 5.98 \times 10^{24}$ kg, $M_m = 7.35 \times 10^{22}$ kg, so

$$K_{moon}/K_{earth} = M_E/M_m = 5.98 \times 10^{24}/7.35 \times 10^{22} = 81.4$$

12.23 HINT: Use $I = \sum m_i r_i^2$. The figure shows a view along the axis that passes through the nitrogen atom and that is perpendicular to the base plane containing the three hydrogen atoms.

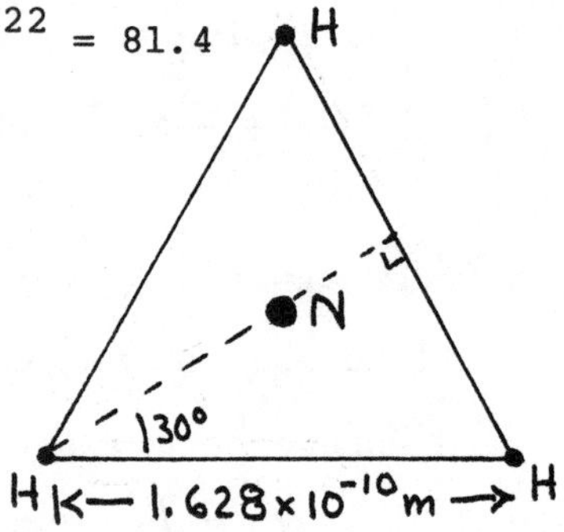

12.33 HINT: The moment of inertia of an object about a given axis is the sum of the moments of inertia of the "pieces" which make up the object. The "pieces" here are the two arms.

12.34 HINT: M is the total mass of the double cone. The distribution of mass relative to the axis AA is the same for the double cone as for a single cone, so the moment of inertia of the double cone is twice that for a single cone of mass M/2. Treat the cone as a stack of circular disks. $I_{cone} = \int dI_{disk}$; $dI_{disk} = \frac{1}{2}dmr^2$

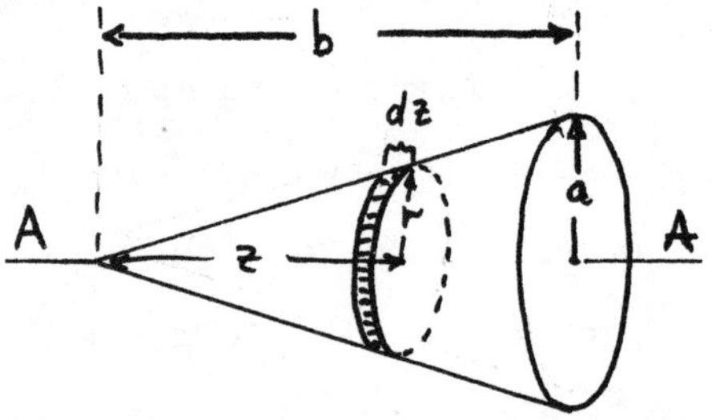

The mass dm of a disk of radius r equals the density times its volume:

$dm = \rho\pi r^2 dz$; The distance z is a convenient variable of integration. From the figure, $r/z = a/b$. You will also have to integrate

$dm = \rho\pi r^2 dz$ to relate the density to the cone mass, M/2.

12.35 HINT: Use the parallel axis theorem.

12.39 The moment of inertia of a strip of width dy and mass dm = M(dy/a) about an axis through its center of mass is

$$dI_{cm} = (1/12)dmb^2$$

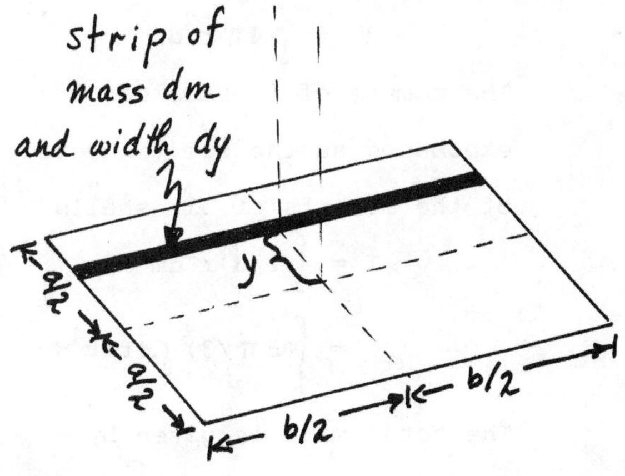

We can use the parallel axis theorem to get the moment of the strip about a parallel axis through the center of mass of the lamina.

$$dI = dI_{cm} + dmy^2$$

Integrating from y = -a/2 to y = a/2 gives the moment of inertia of the complete lamina

$$I = (1/12)b^2 \int dm + \int_{-a/2}^{+a/2} y^2 (M/a)dy$$

$$I = (1/12)b^2M + (M/a)[(a/2)^3 - (-a/2)^3]/3$$

$$I = (1/12)M(a^2 + b^2)$$

12.43 HINT: Here's an opportunity to use your differential calculus. The moment of inertia of a solid sphere about an axis through its center is $2MR^2/5$. The mass of the solid sphere is $M = (4\pi/3)\rho R^3$, where ρ is the mass density. Plug in this result for M to get I for the sphere in terms of ρ and R. You will get $I = aR^5$ where a is a constant. The differential

$$dI = 4aR^4dR$$

represents the moment of inertia of a thin spherical shell of radius R and thickness dR. You can express the mass (dM) of the shell in terms of ρ, R, and dR, and thereby show the desired result $dI = (2/3)dMR^2$, or $I = (2/3)MR^2$

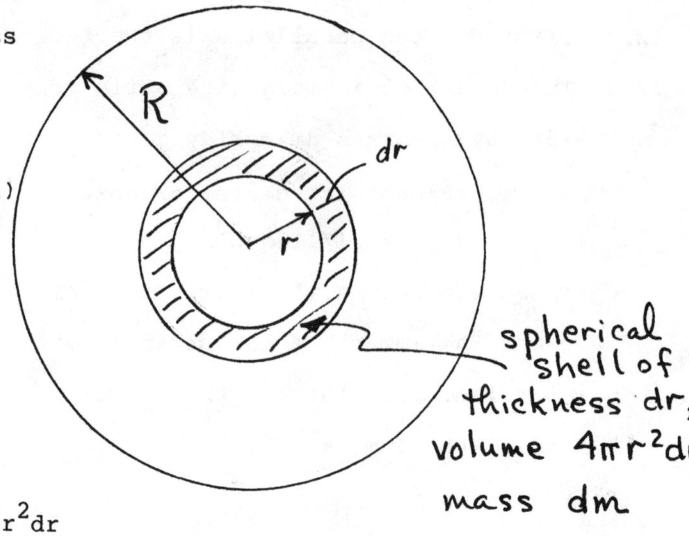

12.44 The spherical shell has a mass

$$dm = \rho 4\pi r^2 dr$$

The moment of inertia can be expressed as the sum (integral) of the moments of all shells

$$I = \int (2/3) r^2 dm$$

$$= \int_0^R (8\pi/3)\,\rho(r) r^4 dr$$

The total mass is given by

$$M = \int dm = \int_0^R 4\pi\rho(r) r^2 dr$$

If we define the dimensionless factor a by $I = aMR^2$ we have

$$aR^2 = I/M = \frac{\int_0^R (8\pi/3)\,\rho(r) r^4 dr}{\int_0^R (4\pi/3)\,\rho(r) r^2 dr}$$

12.45 HINT: Use the result derived in Problem 12.44 to show that a = 0.330.

12.47 HINT: Use $I\alpha = \tau$ and note that the net torque about the hinged end is due solely to the weight of the rod acting at its midpoint.

12.53 HINT: Angular momentum is conserved. The moment of inertia of the final configuration (disk + particle) is 3 times that of the disk alone.

12.56 HINT: Use $I\alpha = \tau$ and note that $\tau = -Fr$ where r is constant and F is a given function of time.

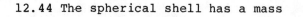

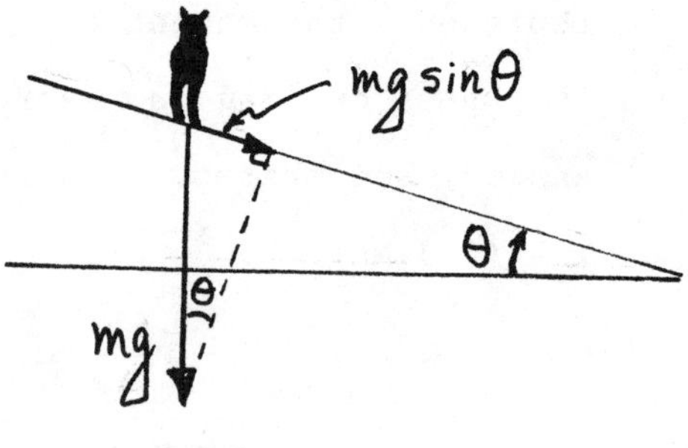

13.2 The component mg sinθ of the horse's weight acts tangentially on the edge of the wheel. It produces a torque $\tau = mg \sin\theta \cdot R$, where R is the radius of the wheel. The power developed is $P = \tau\omega = mg \sin\theta \cdot R\omega$. The walking speed of the horse is $v = R\omega$ so $P = mg \sin\theta\, v$, and

a) $v = P/mg\sin\theta = 746/(800)(9.8)\sin 10^\circ = 0.548$ m/s.

b) With R = 2 m the torque is $\tau = mg \sin\theta \cdot R = (800)(9.8)\sin 10^\circ \cdot 2 = 2720$ N·m

13.4 & 13.5 HINT: Take the moment of inertia of the disk to be $\frac{1}{2}MR^2$.

13.8 HINT: b) The kinetic energy is converted into work at a rate P = 3000 kW. The work performed in time t is Pt. c) $P = \tau\omega$

13.9 b) $\omega = 2\pi/T$; $d\omega/\omega = -(2\pi/T^2)dT/[2\pi/T]$ gives $d\omega = -\omega dT/T$

Setting $dT = 10^{-5}$ s, the yearly change in the length of the day, gives $d\omega$, the yearly change in ω, $d\omega = -(7.29 \times 10^{-5}\ \text{rad/s})(10^{-5}\ \text{s}/86{,}164\ \text{s})$

$$= -8.46 \times 10^{-15}\ \text{rad/s}$$

The minus sign signifies that ω decreases as T increases.

13.12 View the ball as an object rotating about the point of contact P. The total kinetic energy is $K = \frac{1}{2}I_p\,\omega^2$, where ω is the angular velocity about P.

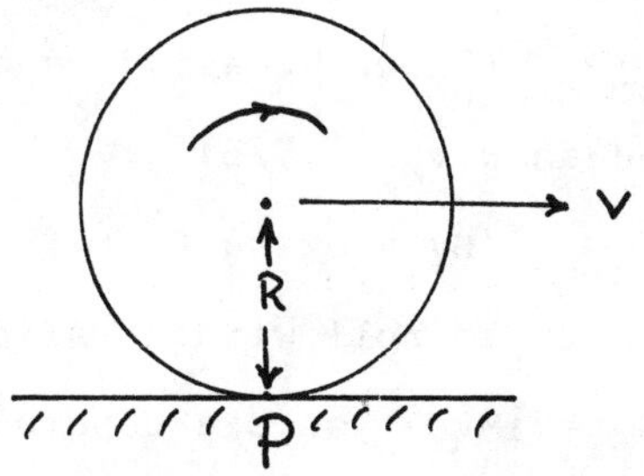

Because the motion is rolly-no-slippy, $v = R\omega$ relates R, ω, and v, the speed of the center of mass. The parallel axis theorem relates I_p and $I_{cm} = (2/3)MR^2$: $I_p = I_{cm} + (2/3)MR^2$;

$K = \frac{1}{2}[(2/3)MR^2 + MR^2]\,\omega^2 = (1/3)Mv^2 + \frac{1}{2}Mv^2$

The second term is the translational kinetic energy of the ball. The first term is the rotational energy of the ball relative to its center of mass. Their ratio is $K_{rot}/K_{trans} = (1/3)Mv^2/\frac{1}{2}Mv^2 = 2/3$.

b) No. In general the ratio $K_{rot}/K_{trans} = I_{cm}/MR^2$. The ratio for a solid sphere is $I_{cm}/MR^2 = 2/5$.

13.15 HINT: There are several key points.

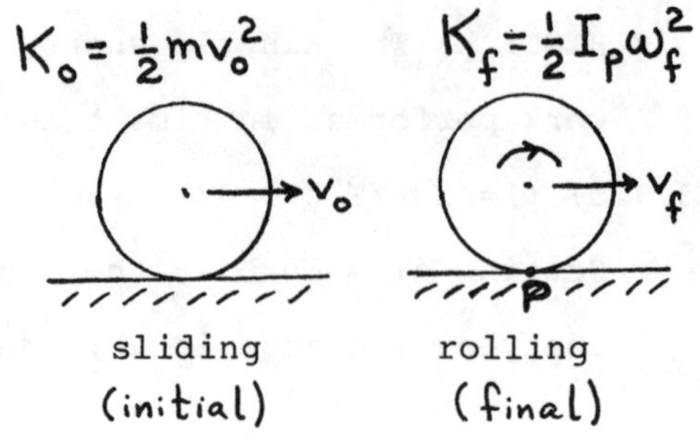

The frictional force F delivers a linear impulse that causes the linear momentum to decrease. $\Delta p = -\int F dt$

The frictional torque FR delivers an angular impulse that changes the angular momentum. $\Delta L = \int FR dt = -R\Delta p$

Δp can be related to mv_f and mv_o. The change in angular momentum equals $I_{cm}\omega_f$ ("without spin" means $\omega_o = 0$). Making use of $R\omega_f = v_f$ you can show $v_o = (7/5)v_f$.

By showing that the kinetic energy (K_f) at the moment the ball starts to roll without slipping is 5/7 of its initial kinetic energy ($K_o = \frac{1}{2}Mv_o^2$) you can conclude that 2/7 of the initial kinetic energy is "lost". To show that $K_f = (5/7)K_o$, note that when the ball rolls without slipping its kinetic energy is related to its moment of inertia about the point of contact (I_p) by $K_f = \frac{1}{2}I_p\omega_f^2$. The parallel axis theorem gives $I_p = (2/5)MR^2 + MR^2 = (7/5)MR^2$.

13.16 HINT: Gravitational potential energy is converted into kinetic energy. Treat the sphere as an object rotating about the instantaneous point of contact. The moment of inertia about the point of contact is $(7/5)MR^2$.

13.17 HINT: Apply conservation of mechanical energy at the three points 1,2,3. The ball begins from rest at 1. $K_1 = 0$, $U_1 = mg \cdot 10R$. The total energy is $E = K + U = 10mgR$. At 2 where

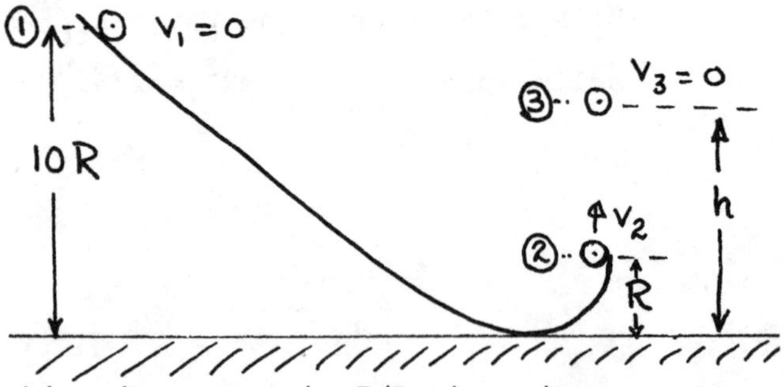

the ball leaves the track its total kinetic energy is 7/5 times its translational kinetic energy; $K_2 = (7/5)\frac{1}{2}mv_2^2$, with the "extra" $(2/5)\frac{1}{2}mv_2^2$ being its rotational energy. At its highest point, 3, its translational energy is zero. Its rotational energy remains unchanged as it rises. Apply energy conservation at 1,2,3, and eliminate $\frac{1}{2}mv_2^2$ to express h in terms of R.

13.19 From Problem 13.15 we know that 2/7 of the initial kinetic energy is "lost" during the transition from pure sliding to rolly-no-slippy motion. This decrease in kinetic energy equals the work done by the frictional force. Thus

$$W_f = Fx = \mu mgx = (2/7)\tfrac{1}{2}mv_o^2$$

$$\mu = v_o^2/7gx = (10)^2/7(9.8)(2) = 0.729$$

13.22 HINT: a) Use energy conservation. The center of mass falls a distance $\frac{1}{2}L\sin\Theta$. b) As viewed from the center of mass the ends move at speeds of $\frac{1}{2}L\omega$ where ω is the angular velocity about the center of mass. The lower end has no vertical component of velocity. Hence, at impact the linear speed of the center of mass (v) equals $\frac{1}{2}L\omega$. The total kinetic energy at impact is $\frac{1}{2}mv^2 + \frac{1}{2}I_{cm}\omega^2$. Treat the ladder as a long thin rod.

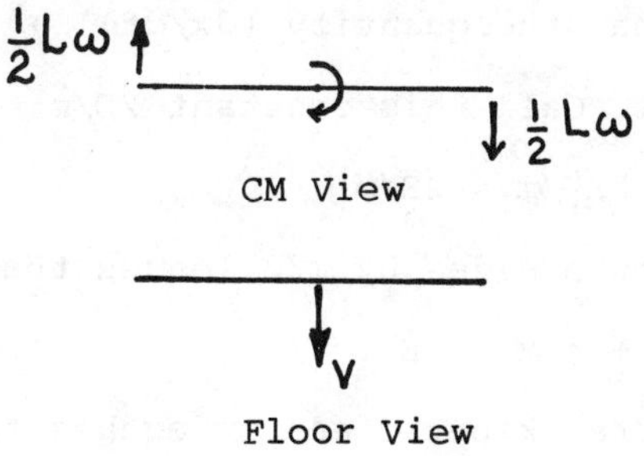

13.26 From the fast top result ($\Omega = \tau/I\omega$)

$$2\pi/T_p = \Omega = \tau/I\omega = \tau T_{spin}/2\pi I$$

$$T_{spin} = (2\pi)^2 I/T_p\tau = (2\pi)^2\ I/T_p mg\ell$$

$$T_{spin} = (2\pi)^2(1.30)/(28)(3.9)(9.8)(0.35) = 0.137\ \text{s}$$

The precession satisfies the fast top conditions:

$$\omega = 2\pi/T_{spin} = 45.9\ \text{rad/s};\quad \Omega = 2\pi/T_p = 2\pi/28 = 0.224\ \text{rad/s}$$

$$\omega \simeq 205\Omega$$

13.27 HINT: Use $\tau = |\Omega \times L|$ and note that the moment arm for the torque produced by the weight is reduced by a factor of $\sin\Theta$ (from ℓ to $\ell\sin\Theta$) when the top is tilted at an angle Θ, rather than at $\Theta = 90^\circ$.

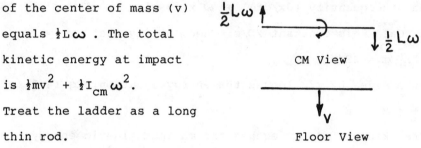

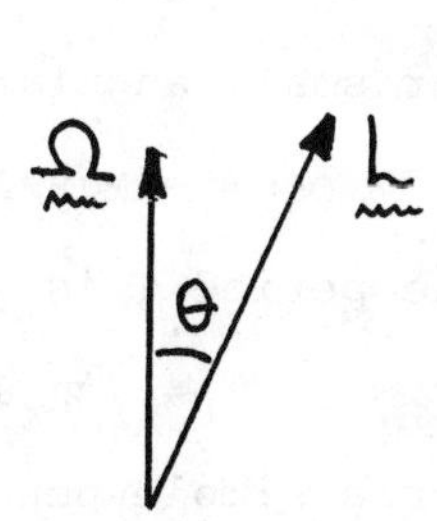

13.28 HINT: The fast top satisfies $\omega/\Omega \gg 1$. The spin energy is $K_{spin} = \frac{1}{2}I\omega^2 = \frac{1}{2}L\omega$. Argue that the maximum decrease in potential energy is approximately equal to the maximum torque exerted by gravity. Using the fast top condition ($\omega = \tau/L$) you can relate $K_{spin}/\Delta U$ to ω/Ω .

14.4 HINT: a) Remember that the second hand requires one minute for a complete revolution.

14.5 & 14.6 HINT: $2\pi/T = \omega = [k/m]^{\frac{1}{2}}$

14.11 Rewrite Eq. 14.4 as $d^2x/dt^2 + \omega^2 x = 0$. Multiply through by $2(dx/dt)$ to get $2(dx/dt)(d^2x/dt^2) + \omega^2 \cdot 2x(dx/dt) = 0$

Note the identities: $d/dt[(dx/dt)^2] = 2(dx/dt)(d^2x/dt^2)$

$$d/dt[x^2] = 2x(dx/dt)$$

The equation becomes $d/dt[(dx/dt)^2] + d/dt[\omega^2 x^2] = 0$

This shows that the time derivative of $(dx/dt)^2 + \omega^2 x^2$ is zero.

It follows that the quantity $(dx/dt)^2 + \omega^2 x^2 = v^2 + \omega^2 x^2$ is a constant. Call this constant $2E/m$. Then, noting that $\omega^2 = k/m$,

$$v^2 + kx^2/m = 2E/m$$

Multiplying both sides by $m/2$ leaves the energy equation, Eq 14.7.

$$\tfrac{1}{2}mv^2 + \tfrac{1}{2}kx^2 = E$$

14.13 HINT: The total kinetic energy equals the maximum kinetic energy.

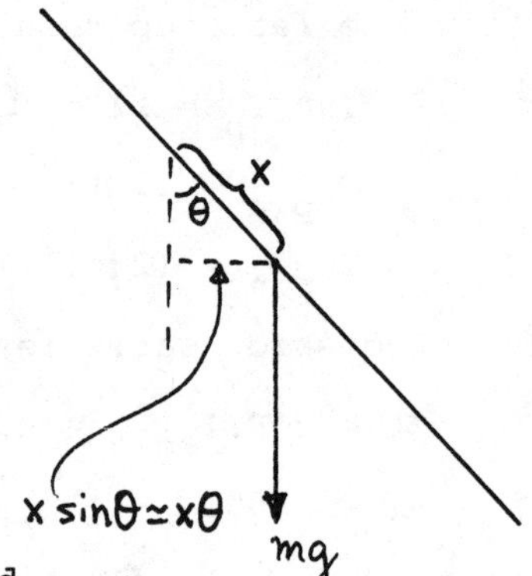

14.19 Let the axis about which the stick oscillates be a distance x from the center. The parallel axis theorem gives the moment of inertia about that axis as

$$I = Mx^2 + (1/12)ML^2$$

The equation of motion is

$$I\alpha = \tau = -Mgx \sin\theta$$

For small angular displacements $\sin\theta \simeq \theta$ and

$$\alpha = -(Mgx/I)\theta = -\omega^2\theta = -(2\pi/T)^2\theta$$

The period T is given by

$$T = 2\pi[I/Mgx]^{\frac{1}{2}} = 2\pi[(x^2 + L^2/12)/gx]^{\frac{1}{2}}$$

For a simple pendulum of length ℓ, $T = 2\pi[\ell/g]^{\frac{1}{2}}$ so the length of a simple pendulum having the same period is $\ell = x + L^2/12x$

For an axis 20 cm from the end, $x = 0.3$ m and $\ell = 0.3 + 1^2/12(.3) = 0.581$ m

The period of the pendulum is $T = 2\pi[0.581/9.80]^{\frac{1}{2}} = 1.53$ s

14.20 HINT: From the solution of 14.19

$$T^2 = C\{x + L^2/12x\},$$ where C is a constant.

To find the value of x for which T (& thus T^2) is a minimum, impose the condition $d[T^2]/dx = 0$.

14.23 HINT: Draw a figure and express x in terms of ℓ and θ. Use the small-angle ($\theta << 1$) form of the relation to replace θ by a combination of x and ℓ.

14.26 HINT: The spring constant k is <u>inversely</u> proportional to the length of the spring. What happens to k when L is reduced to $\frac{1}{2}L$?

14.28 Method 1. Take natural logs to rewrite $T = 2\pi[\ell/g]^{\frac{1}{2}}$ as $\ln T = \frac{1}{2}\ln\ell + C$ where C is a constant. Recall $d(\ln x) = dx/x$ and form the differential of both sides to get $(dT/T) = \frac{1}{2}(d\ell/\ell)$. Identify dT as ΔT and $d\ell$ as $\Delta\ell$.

Method 2. Form the differential of $T = 2\pi[\ell/g]^{\frac{1}{2}}$, then divide the result by the original equation to get $(dT/T) = \frac{1}{2}(d\ell/\ell)$ as before.

14.31 HINT: Use the result for T^2 derived in 14.19, with s = x. See also the hint for 14.20.

14.33 $v = 2\pi A/T = 2\pi(9.3 \times 10^7 \text{ mi})/1 \text{ yr} = 5.84 \times 10^8 \text{ mi/yr}$

$a = v^2/A = (5.84 \times 10^8)^2/9.3 \times 10^7 = 3.67 \times 10^9 \text{ mi/yr}^2$

14.39 a) $T = 2\pi[\ell/g]^{\frac{1}{2}} = 2\pi \text{ s} \Rightarrow [\ell/g]^{\frac{1}{2}} = 1 \text{ s} \Rightarrow \ell = 9.80 \text{ m}$

The characteristic decay time is $\tau = 2m/\gamma = 2(70 \text{ kg})/(0.01 \text{ kg/s}) = 14{,}000 \text{ s}$

Comparing τ with the period T = 6.28 s shows that $T << \tau$ - the motion is underdamped. For time intervals much less than τ we can treat the motion as undamped.

b) For an undamped oscillator energy is conserved and the maximum kinetic energy equals the maximum potential energy. The maximum speed follows from energy conservation.

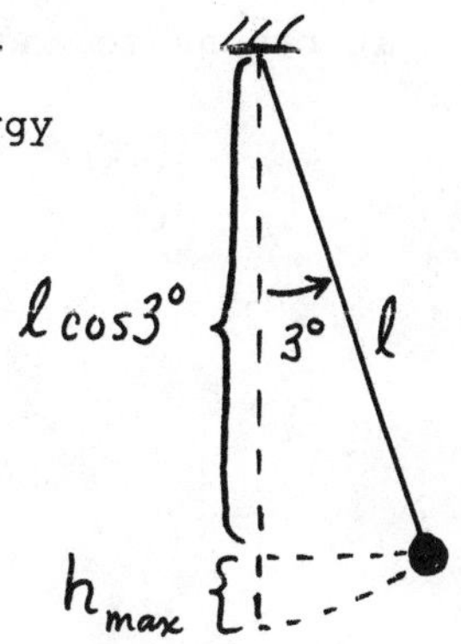

14.39 (cont) $(\frac{1}{2}mv^2)_{max} = mgh_{max} = mg\ell(1 - \cos 3^o)$

$$v_{max} = [2g\ell(1 - \cos 3^o)]^{\frac{1}{2}} = [2(9.8)(9.8)(0.00137)]^{\frac{1}{2}}$$

$$= 0.513 \text{ m/s}$$

c) $K_{max} = \frac{1}{2}mv^2_{max} = \frac{1}{2}(70)(0.513)^2 = 9.21$ J

d) The maximum angular displacement, the maximum linear displacement, and the maximum speed "decay" exponentially with time according to $e^{-t/\tau}$ [See Eq 14.41, for example]. The maximum kinetic energy is proportional to the square of the maximum speed and so decays exponentially with time according to $(e^{-t/\tau})^2 = e^{-2t/\tau}$. Specifically,

$$K_{max}(t) = K_{max}(0)e^{-2t/\tau}$$

With

$$K_{max}(0) = 9.21 \text{ J}$$

$$t = 2 \text{ hr} = 7200 \text{ s},$$

$$2t/\tau = 2(7200 \text{ s})/14{,}000 \text{ s} = 1.02857$$

$$K_{max}(2 \text{ hr}) = (9.21 \text{ J})e^{-1.02857} = 3.29 \text{ J}$$

e) From $P = -\gamma v^2 = (2\gamma/m)(\frac{1}{2}mv^2)$

$$P_{max}(t) = -(2\gamma/m)K_{max}(t)$$

$$P_{max}(0) = -(2\gamma/m)K_{max}(0) = -([2(0.01)/70])(9.21) \text{ J/s}$$

$$= -2.63 \times 10^{-3} \text{ W}$$

f) $P_{max}(t) = -(2\gamma/m)K_{max}(t)$; $P_{max}(2 \text{ hr}) = -(2\gamma/m)K_{max}(2 \text{ hr})$

$$P_{max}(2 \text{ hr}) = -([2(0.01)/70])(3.29) \text{ J/s} = -9.40 \times 10^{-4} \text{ W}$$

14.44 HINT: Small damping requires $\omega\tau >> 1$. The resonance and low-frequency expressions for A are given by equations 14.51 and 14.53. Form the ratio of A at resonance and A in the low-frequency limit.

15.4 ρ = mass/volume = mass/$(4\pi r^3/3)$: For the earth, mass = 5.98×10^{24} kg and r = 6.37×10^6 m. $\rho = 5.98 \times 10^{24}$ kg/$(4\pi[6.37 \times 10^6\ \text{m}]^3/3)$

$$\rho = 5.52 \times 10^3\ \text{kg/m}^3$$

Mercury, Venus, and Mars have densities close to that of the earth, which is one reason they are called earthlike. The Jovian planets all have densities significantly smaller than that of the earth, and closely comparable to the density of Jupiter. For Jupiter,

$$\rho = 1.90 \times 10^{27}\ \text{kg}/(4\pi[6.97 \times 10^7\ \text{m}]^3/3) = 1.34 \times 10^3\ \text{kg/m}^3$$

15.8 a) The pressure equals the weight (mg) divided by the area on which it acts.

$$P = mg/A = (60\ \text{kg})(9.80\ \text{m/s}^2)/(0.1\ \text{m})(3 \times 10^{-3}\ \text{m})$$

$$= 1.96 \times 10^6\ \text{N/m}^2 = 1.96 \times 10^6\ \text{Pa}$$

b) Multiply by the conversion factor 1 atm/1.01×10^5 Pa to express the pressure in atmospheres.

$$P = 1.96 \times 10^6\ \text{Pa}\cdot(1\ \text{atm}/1.01 \times 10^5\ \text{Pa}) = 19.4\ \text{atm}$$

15.9 HINT: Use the Pythagorean theorem to find $\Delta \ell$.

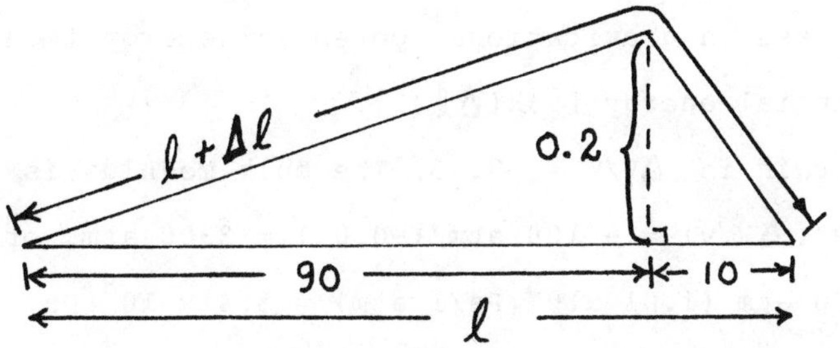

15.13 HINT: The atmospheric pressure equals the weight of the atmosphere divided by the surface area of the earth: $P = mg/4\pi r_E^2$. Knowing P, g, and r_E enables you to calculate the mass of the atmosphere.

15.23 HINT: The basic stress-strain relation (Eq 15.9) gives the increase in length,

$$\Delta \ell = F_t \ell / YA$$

F_t is the tensile force exerted on the cord by the bob. To evaluate F_t, consider the forces acting on the bob at the bottom of its swing. By Newton's third law the cord exerts an upward force F_t. The net force on the bob is F_t - mg. This net force gives the bob a radial (centripetal) acceleration $v^2/(\ell + \Delta\ell)$. Use Newton's second law and the conservation of mechanical energy (expressing the conversion of gravitational potential energy into kinetic energy) to get two equations involving mv^2. By eliminating mv^2 you can show that F_t = 3mg, which gives the sought-for result $\Delta\ell = 3mg\,\ell/YA$.

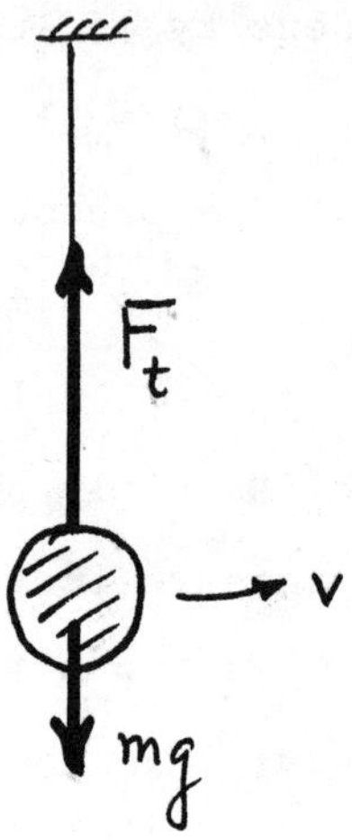

15.24 HINT: The decrease in gravitational potential energy [mgh] is stored as elastic potential energy [$\frac{1}{2}k(\Delta\ell)^2$].

15.27 The volume strain is $\Delta V/V = -0.03$. The bulk modulus is

$$B = -P/(\Delta V/V) = -100 \text{ atm}/(-0.03) = 3300 \text{ atm, or}$$

$$B = 3300 \text{ atm } (1.01 \times 10^5 \text{ Pa/1 atm}) = 3.4 \times 10^8 \text{ Pa}$$

The compressibility is $K = 1/B = 3.0 \times 10^{-9}$/Pa

15.29 HINT: Use $m = \rho V$ = constant. Differentiate to find the relation between differentials: $\rho dV + V d\rho = 0 \Rightarrow (d\rho/\rho) = -(dV/V)$, or

$$\Delta\rho/\rho = -(\Delta V/V)$$

16.2

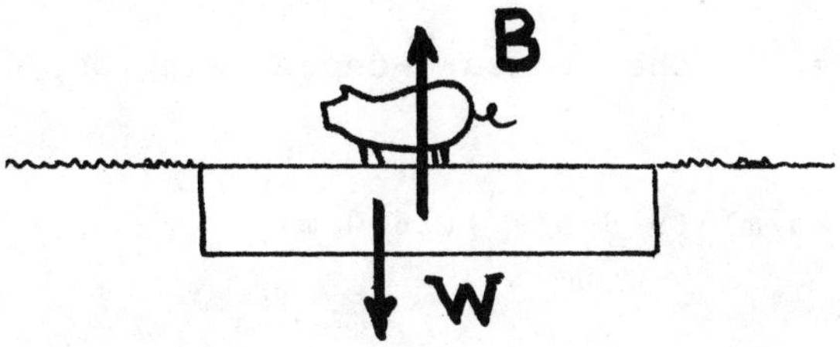

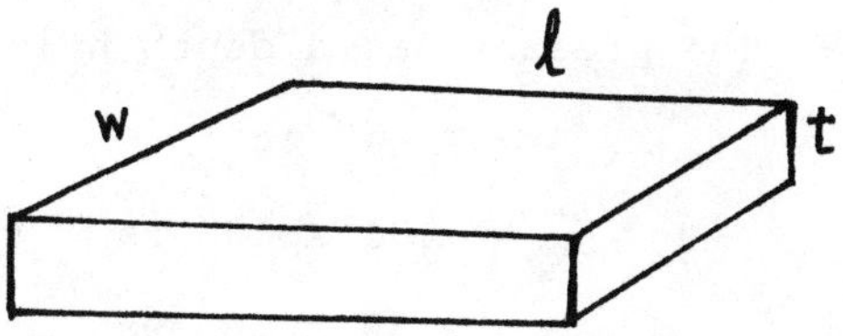

The submerged volume $V = \ell wt$ displaces water. The mass of the displaced water is $m = \rho_{H_2O}V = \rho_{H_2O}\ell wt$. The buoyant force equals the weight of this mass.

$$B = mg = \rho_{H_2O}\ell wtg$$

The combined weight of the pig and board is

$$W = 425\ \text{N} + \rho_{wood}\ell wtg$$

The board floats so B = W. Solving B = W for the width w gives

$$w = 425/[(10^3-600)(3)(0.15)(9.8)] = 0.24\ \text{m} = 24\ \text{cm}$$

16.5 HINT: Because the cube's density is intermediate between that of the Cognac and the Curacao it will float between these two layers. Equate the weight of the cube to the sum of the buoyant forces exerted by the Cognac and Curacao.

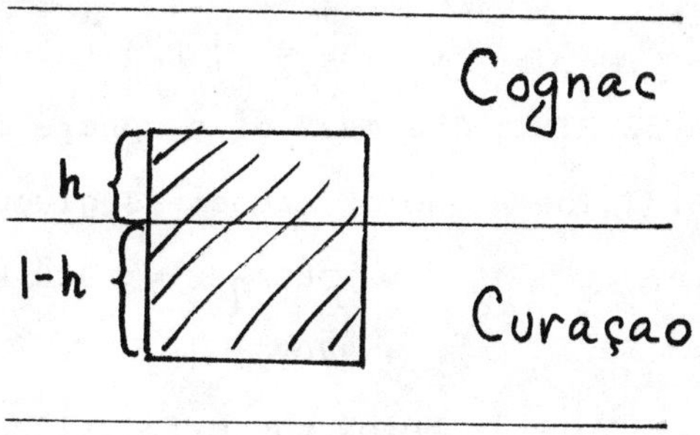

16.7 HINT: The buoyant force is

$$B = \rho_{air}Vg$$

The weight of the 100 kg blimp and its helium is

$$W = \rho_{He}Vg + (100\ \text{kg})g$$

16.9 The pressure at a depth h is given by the pressure-depth relation,

$$P = P_o + \rho gh$$
$$= 1.01 \times 10^5 \text{ Pa} + (950 \text{ kg/m}^3)(9.8 \text{ m/s}^2)(1600 \text{ m})$$
$$= 1.50 \times 10^7 \text{ Pa}$$

The pressure $P_o = 0.01 \times 10^7$ Pa is due to the air.

16.15 HINT: $P = \rho_{air} gh$ gives the pressure at the bottom of an "ocean" of air of depth h.

16.16 From the equation of continuity, Eq 16.8,

$$A_1 u_1 = A_2 u_2$$

Let u_1 denote the flow speed in the nozzle and let u_2 denote the flow speed in the hose. Then

$$u_1 = A_2 u_2 / A_1 = (\pi d_2^2/4) u_2 / (\pi d_1^2/4)$$
$$= (5/8)^2 (0.43 \text{ m/s})/(1/4)^2 = 2.69 \text{ m/s}$$

16.20 The discharge rate is given by Eq 16.9

$$Q = Au = (\pi/4)(0.0512 \text{ m})^2 (1.15 \text{ m/s}) = 2.37 \times 10^{-3} \text{ m}^3/\text{s}$$

16.32 HINT: The mass of a sphere of radius r and density ρ is $4\pi r^3 \rho/3$.

16.37 The Reynolds number is given by Eq 16.24,

$$R = \rho uL/\eta = (900 \text{ kg/m}^3)(3 \text{ m/s})(36 \times 10^{-3} \text{ m})/0.10 \text{ kg/m}\cdot\text{s}$$
$$= 970$$

This is below the critical Reynolds number range (1,000 - 10,000) so the flow is laminar.

17.4 We can compare $\psi = 2.0 \sin(2.11x - 3.62t)$ cm with Eq 17.12 which gives one of the standard forms of a sinusoidal waveform,

$$\psi = A \sin(kx - \omega t) \qquad \text{Eq 17.12}$$

We see that $A = 2.0$ cm, $k = 2.11$ rad/m, $\omega = 3.62$ rad/s. The wavelength is given by $\lambda = 2\pi/k = 2\pi\,\text{rad}/(2.11\ \text{rad/m}) = 2.98$ m.

The wavespeed is $v = \omega/k = 3.62\ \text{rad/s}/(2.11\ \text{rad/m}) = 1.72$ m/s

17.17 HINT: Eq 17.17 relates phase difference and spatial separation.

17.20 HINT: The smallest separation that gives a 180° phase difference is half of one wavelength. Take the speed of sound to be 344 m/s.

17.23 HINT: Let $\varphi = kx + \pi/4$ so that $y_1 = A \sin(\varphi - \pi/4)$; $y_2 = A \sin(\varphi + \pi/4)$. Then use the addition theorems for $\sin(\alpha \pm \beta)$ [Appendix 4] to show $y_1 + y_2 = \sqrt{2}\,A\sin\varphi$, which is a sinusoidal wave with an amplitude $\sqrt{2}\,A$.

17.25 HINT: The beat frequency equals the difference between the frequencies of the two forks. Knowing the frequency of one fork and the beat frequency allows for two possible frequencies of the second fork. The additional datum allows you to decide between these two possible frequencies.

17.26 The two radian frequencies are $\omega_2 = 4330$rad/s and $\omega_1 = 4320$ rad/s. The beat frequency is given by Eq 17.26, $\nu_{beat} = \nu_2 - \nu_1$.

$$\nu_{beat} = \nu_2 - \nu_1 = (\omega_2 - \omega_1)/2\pi = 10/2\pi = 1.59\ \text{Hz}$$

17.28 HINT: See the solution presented for problem 17.31a. The radar signal travels at the speed of light, 3×10^8 m/s. Because the car moves toward the radar unit the "fuzz-buster" receives a higher frequency than the frequency broadcast by the radar source.

17.31

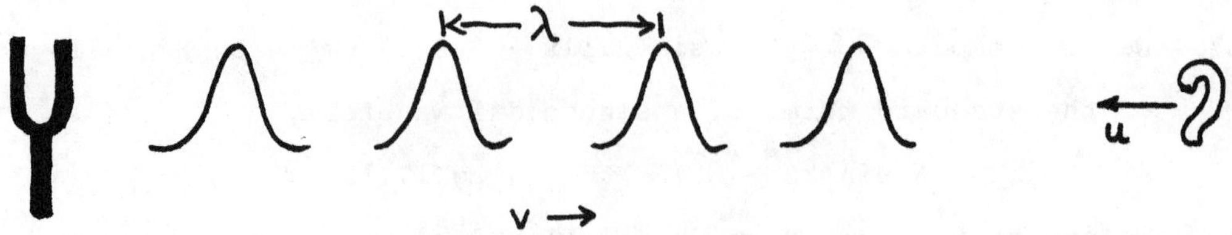

a) The figure shows a source that generates wave crests separated by a distance λ. The source period is T_o and the wave speed is v. The kinematic relation between λ, v, and T_o is

$$\lambda = vT_o$$

The observer moving toward the source at speed u encounters wave crests at a relative speed of v + u. The crest-to-crest spacing is λ, the same as it is for an observer at rest. If T denotes the period of the waves for the moving observer then the kinematic relation for the moving observer is

$$\lambda = (v + u)T$$

Equating these two expressions for λ, and solving for T_o/T gives

$$vT_o = (v + u)T \quad \Rightarrow \quad T_o/T = (v + u)/v = 1 + u/v$$

In terms of frequencies,

$\nu_o = 1/T_o$ frequency of source

$\nu = 1/T$ frequency perceived by moving observer

Substituting $\nu/\nu_o = T_o/T$ gives

$$\nu = (1 + u/v)\nu_o$$

The Doppler shift is $\Delta\nu = \nu - \nu_o$

$$\Delta\nu = \nu - \nu_o = (1 + u/v)\nu_o - \nu_o = (u/v)\nu_o$$

b) Using the result of part a) gives

$$\Delta\nu = (u/v)\nu_o = (60/760)256 \text{ Hz} = 20.2 \text{ Hz}$$

Using 17.32 gives

$$\Delta\nu = [256 \text{ Hz}/(1 - 60/760)] - 256 \text{ Hz} = 21.9 \text{ Hz}$$

When the ratio u/v = (speed of the observer/wave speed) is much smaller than unity the result given by Eq 17.32 reduces to the result derived in part a). To show this, use the binomial expansion [Appendix 4] to expand $[1 - u/v]^{-1}$ in Eq 17.32 . This gives $\Delta\nu = \nu - \nu_o = \nu_o[(u/v) + (u/v)^2 + \ldots]$. When $(u/v) \ll 1$ all but the first term can be ignored, giving $\Delta\nu \simeq \nu_o(u/v)$, which is the result derived in part a) for a moving observer.

18.4 Equation 18.11 can be rearranged to express the tension T in terms of M, L, and the wave speed v,

$$T = (M/L)v^2$$

The wave travels 10 m in 0.11 s so its speed is

$$v = 10 \text{ m}/0.11 \text{ s} = 90.9 \text{ m/s}$$

The tension is

$$T = (0.75 \text{ kg}/10 \text{ m})(90.9 \text{ m/s})^2 = 620 \text{ N}$$

18.11 The speed of longitudinal sound waves is given by Eq 18.26

$$v = [B/\rho]^{\frac{1}{2}} = [2.40 \times 10^9/1.345 \times 10^4]^{\frac{1}{2}} = 422 \text{ m/s}$$

18.15 HINT: Use eq 18.24 and 18.25 to show that $v_{long}/v_{tran} = [4/3 + B/\mu]^{\frac{1}{2}}$

The minimum value of the ratio of wave speeds occurs for B = 0.

18.18 HINT: The waves travel equal distances so the ratio of their speeds v_{long}/v_{tran} is equal to t_{tran}/t_{long}, the ratio of transit times. Equations 18.24 and 18.25 let you relate v_{long}/v_{tran} to B/μ.

18.22 $\beta = 10 \log_{10}(I_2/I_1) \text{ dB} = 10 \log(6 \times 10^{-12}/10^{-12}) \text{ dB}$

$= 10 \log_{10}(6) \text{ dB} = 7.78 \text{ dB}$

The change from 6×10^{-12} W/m² to 10^{-11} W/m² gives

$$\beta = 10 \log_{10}(10^{-11}/6 \times 10^{-12}) \text{ dB} = 2.22 \text{ dB}$$

18.29 HINT: The intensity equals the power per unit area. The intensity in part b) is twice that of part a). You should explain why this is so.

18.32 HINT: The intensity can be expressed as A/r^2 where r is the distance from the source and A is a constant (whose value you need not determine). By forming the ratio of intensities you can determine the value of $(r + 1)/r$ and thus r.

18.36 HINT: Equation 18.50 will let you estimate the fundamental frequency. The author obtained a value of approximately 20 kHz for the fundamental frequency of his skull.

18.37 HINT: The speed of shallow water waves is given by Eq 18.20, $v = [gh]^{\frac{1}{2}}$.

19.1

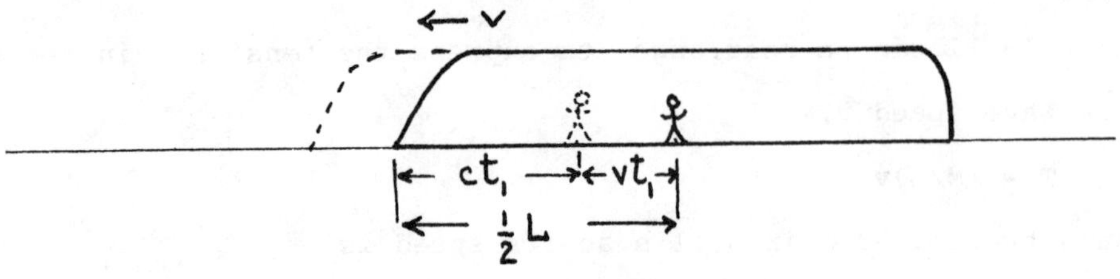

Let t_1 be the time required for the light to travel from the front of the train to the president. Let t_2 denote the time for the light to travel to him from the rear of the train. From the figure we see that the president moves forward a distance vt_1 during the time t_1. The oncoming light travels a distance ct_1. The sum of these two distances is $\frac{1}{2}L$, half the length of the train [Remember, the president stands at the middle of the train]. Thus,

$$\tfrac{1}{2}L = ct_1 + vt_1 \;; \quad t_1 = \tfrac{1}{2}L/(c + v)$$

The light from the rear of the train travels a distance $\frac{1}{2}L + vt_2$ before reaching the president.

$$\tfrac{1}{2}L + vt_2 = ct_2 \;; \quad t_2 = \tfrac{1}{2}L/(c - v)$$

The difference between light-flight times is

$$t_2 - t_1 = \tfrac{1}{2}L[1/(c-v) - 1/(c+v)] = Lv/(c^2-v^2)$$

Because $c \gg v$ we can replace $c^2 - v^2$ by c^2 in the denominator.

$$t_2 - t_1 \simeq Lv/c^2$$

Inserting $L = 180$ m, $v = 40$ m/s, $c = 3 \times 10^8$ m/s gives

$$t_2 - t_1 = (180)(40)/(3 \times 10^8)^2 = 8 \times 10^{-14} \text{ s}$$

19.7 HINT: This problem asks that you first obtain a solution assuming Galilean relativity applies. You will find it helpful to do this for ALL special relativity problems involving relative motion.

If u and u' are the velocities of the jets relative to the galaxy their relative velocity is $v = u - u'$ according to Galilean relativity. The Einstein velocity addition formula is

$$u = (u' + v)/(1 + u'v/c^2)$$

You can verify [or argue "by symmetry"] that

$$v = (u - u')/(1 - uu'/c^2)$$

Notice that this special relativity relation reduces to the Galilean result $v = u - u'$ in the low velocity limit.

19.9 HINT: Let

v_{EK} = velocity of Enterprise relative to Klingons

$v_{E\oplus}$ = velocity of Enterprise relative to Earth

$v_{\oplus K}$ = velocity of Earth relative to Klingons

The Galilean velocity addition relation, $v_{EK} = v_{E\oplus} + v_{\oplus K}$ is replaced in special relativity by the corresponding Einstein relation

$$v_{EK} = (v_{E\oplus} + v_{\oplus K})/(1 + v_{E\oplus}v_{\oplus K}/c^2)$$

Note that you are dealing with velocities. You can check the signs you assign to velocities by using the Galilean velocity addition relation as a "common sense" check.

19.11 Consider a rod, at rest in the primed frame of reference. The endpoints of the rod are located at x'_2 and x'_1 so that the length of the rod in its rest frame is $\Delta x' = x'_2 - x'_1$. An observer in the unprimed frame makes measurements as the rod moves past at speed v. She relates her data (one endpoint is at x_2 at time t_2; the other is at x_1 at time t_1) to the primed frame measurements via the Lorentz transformation.

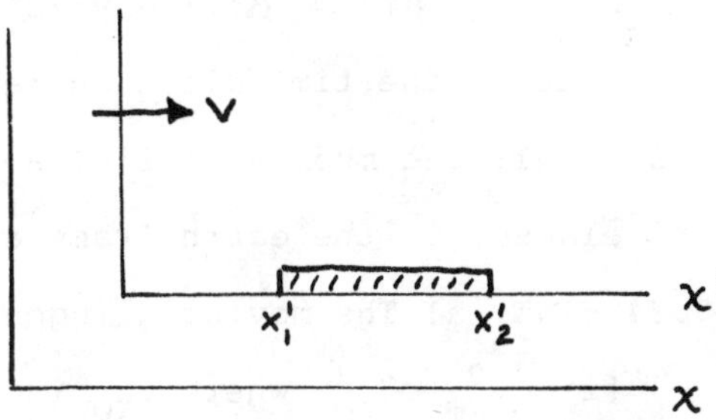

$$x'_2 = (x_2 - vt_2)/[1 - v^2/c^2]^{\frac{1}{2}}$$
$$x'_1 = (x_1 - vt_1)/[1 - v^2/c^2]^{\frac{1}{2}}$$

Subtracting gives

$$x'_2 - x'_1 = [(x_2 - x_1) - v(t_2 - t_1)]/[1 - v^2/c^2]^{\frac{1}{2}}$$

or

$$\Delta x' = (\Delta x - v\Delta t)/[1 - v^2/c^2]^{\frac{1}{2}}$$

In making a length measurement you must locate the endpoints simultaneously. Thus the observer in the unprimed frame measures x_2 and x_1 at the same moment - which means $t_2 = t_1$, $\Delta t = 0$. With $\Delta t = 0$,

$$\Delta x = \Delta x'[1 - v^2/c^2]^{\frac{1}{2}},$$ which states

$$\begin{pmatrix}\text{length of}\\ \text{rod moving}\\ \text{at speed v}\end{pmatrix} = \begin{pmatrix}\text{length of}\\ \text{rod at}\\ \text{rest}\end{pmatrix}\cdot[1 - v^2/c^2]^{\frac{1}{2}}$$

19.12 The binomial expansion [Appendix 4] for $(1 - x)^{\frac{1}{2}}$ is

$$(1 - x)^{\frac{1}{2}} = 1 - \tfrac{1}{2}x - (1/8)x^2 - \ldots..$$

Using this to expand $L = L'[1 - v^2/c^2]^{\frac{1}{2}}$ gives

$$L = L'(1 - \tfrac{1}{2}v^2/c^2 - (1/8)v^4/c^4 - \ldots.)$$

If $v/c \ll 1$ we can ignore all but the first two terms to get

$$L = L'(1 - \tfrac{1}{2}v^2/c^2)$$

19.17 Let $\Delta t'$ denote the time to make a round trip in the rocket frame of reference. Let Δt denote the round trip time for an observer on a planet who sees the rocket move past at speed v. The light speed is c in both frames. The light paths for the two observers are shown in the figure. The Pythagorean formula relates $\Delta t'$ and Δt.

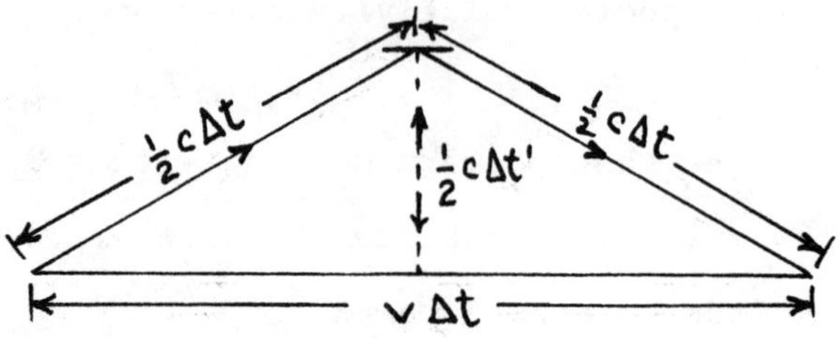

$$(\tfrac{1}{2}c\,\Delta t')^2 = (\tfrac{1}{2}c\,\Delta t)^2 - (\tfrac{1}{2}v\,\Delta t)^2$$

$$\Delta t' = \Delta t[1 - v^2/c^2]^{\frac{1}{2}}$$

This is the time dilation relation: The moving clock runs slow.

19.22 HINT: The muon acts like a moving clock. In its rest frame a time of 2.2 μs elapse. In the earth frame a longer ("dilated") time elapses.

19.24 HINT: a) The moving Klingon clock runs slow by the factor $[1 - v_{K\oplus}^2/c^2]^{\frac{1}{2}}$, where $v_{K\oplus}$ is the speed of the Klingons relative to Earth.

b) Reason from the Klingon frame using the relation

distance traveled = speed x time

c) Reason from the Klingon frame, using the relation

initial separation/relative speed = time to overtake

20.2 $p/mc = (v/c)/(1 - v^2/c^2)^{\frac{1}{2}} = 10 \Rightarrow (v/c) = (100/101)^{\frac{1}{2}} = 0.995$

20.4 HINT: a) Use Eq 20.4 to show

$$F/m = 2x/[t^2 - (x/c)^2]$$

Note that $x = 4.3$ ly $\Rightarrow x/c = 4.3$ yr.

b) Use Eq 20.3 with the value of F/m you obtain in part a).

20.12 a) $K/mc^2 = \frac{1}{2}mv^2/mc^2 = \frac{1}{2}(v/c)^2 = \frac{1}{2}(1/3)^2 = 0.0556$

b) $K/mc^2 = (1 - v^2/c^2)^{-\frac{1}{2}} - 1 = (1 - 1/9)^{-\frac{1}{2}} - 1 = 0.0607$

20.19 HINT: Use $E^2 = (pc)^2 + E_0^2$. The rest energy of the electron is 0.511 MeV.

20.23 From Eq 20.1, $\mathbf{p} = m\mathbf{v}/(1 - v^2/c^2)^{\frac{1}{2}}$,

and Eq 20.12, $E = mc^2/(1 - v^2/c^2)^{\frac{1}{2}}$, we form the ratio $\mathbf{p}/E$ to get

$$\mathbf{p}/E = \mathbf{v}/c^2 \Rightarrow \mathbf{v} = \mathbf{p}c^2/E.$$

In the nonrelativistic limit, $v \ll c$, $E \rightarrow mc^2$ and $\mathbf{v} \rightarrow \mathbf{p}c^2/mc^2 = \mathbf{p}/m$.

20.25 HINT: Use $E = m_\nu c^2/(1 - v^2/c^2)^{\frac{1}{2}}$. With $E \gg m_\nu c^2$ the speed v is so close to c that you can use the approximation

$$1 - v^2/c^2 = (1 + v/c)(1 - v/c) \simeq 2(1 - v/c)$$

20.28 HINT: The two colliding protons have equal energies so the total energy before the interaction is $2m_p c^2/(1 - v^2/c^2)^{\frac{1}{2}}$. Recall [p 386] that $(v/c)^2 = 3/4$ for the colliding protons. The final energy of the system is the rest mass energy of 3 protons and 1 antiproton.

20.29 We present two solutions. The first solution takes advantage of the fact that the two colliding particles have equal rest masses. The second solution is much more general. It applies to any 2-particle interaction in which one particle is initially at rest.

SOLUTION I. Let v_{lab} denote the speed of the incident proton in the lab frame. In the center of momentum (CM) frame the two protons approach each other at the same speed, v. The CM speed v is related to the lab speed by the Einstein addition formula

$$v_{lab} = (v + v)/(1 + v^2/c^2)$$

which can be expressed in the more convenient form

$$v_{lab}/c = \beta_{lab} = 2\beta/(1 + \beta^2); \qquad \beta = v/c$$

In the CM frame the two colliding protons have equal energies. The total energy before the interaction is therefore

$$E_i = 2E_o/(1 - \beta^2)^{\frac{1}{2}}$$

where E_o is the proton rest mass energy (E_o = 938.25 MeV). The final state of the system consists of two protons and a pair of pions. At threshhold these four particles will be at rest in the CM frame. The total energy in the CM frame is thus

$$E_f = 2E_o + 2E_\pi$$

where E_π = 139.58 MeV is the pion rest mass energy. Equating E_i and E_f gives an equation that can be solved for β.

$$E_i = 2E_o/(1 - \beta^2)^{\frac{1}{2}} = 2E_o + 2E_\pi = E_f$$

$$\beta^2 = 1 - (E_o/E_o + E_\pi)^2$$

$$\beta = [1 - (938.25/938.25+139.58)^2]^{\frac{1}{2}} = 0.49217$$

The value of $\beta_{lab} = v_{lab}/c$ is

$$\beta_{lab} = 2\beta/(1 + \beta^2) = 2(0.49217)/(1.24223) = 0.79240$$

The kinetic energy of the proton in the lab frame is

$$K = E_o/(1 - \beta^2_{lab})^{\frac{1}{2}} - E_o = 0.6393E_o$$

$$= 0.6393(938.25 \text{ MeV}) = 600 \text{ MeV}$$

This solution is relatively simple because the two colliding particles have equal rest masses. As a result we are able to solve the problem by making use of the Einstein velocity addition formula. It was not necessary to make explicit use of the conservation of linear momentum. We now develop a more general solution and show how it leads to the same result for this particular problem.

SOLUTION II. The figure shows an interaction in which an incident particle collides with a target particle. The target particle is at rest in the lab frame of reference. When the two particles collide we view them as a single "complex" particle. The complex particle then breaks up into some final configuration of particles.

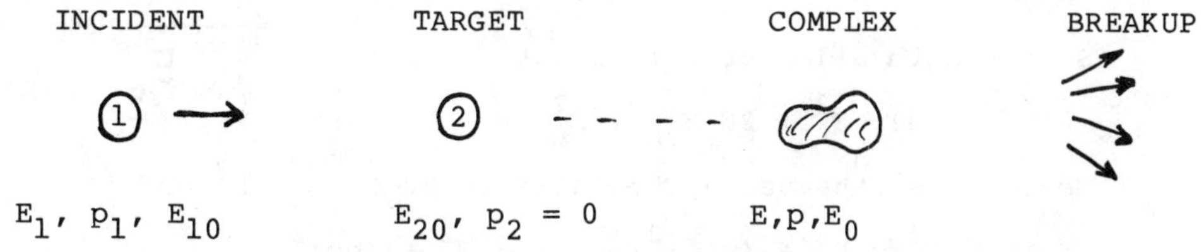

In this particular problem the complex particle breaks up into two protons and two pions. The complex particle idea makes it easy to apply the conservation of energy and linear momentum. We use the following symbols:

INCIDENT PARTICLE:

E_{10}, rest mass energy; E_1, total energy; p_1, linear momentum

TARGET PARTICLE:

$E_{20} = E_2$, rest energy = total energy; $p_2 = 0$, linear momentum

COMPLEX PARTICLE:

E_0, rest mass energy; E, total energy; p, linear momentum

Note: E_0 is the rest mass energy of the complex particle - the energy we would find it possessed if we were riding along with it. Because energy is conserved E_0 is also the total rest mass energy of the particles which result when the complex particle breaks up.

We want to express the <u>kinetic energy</u> of the incident particle

$$K = E_1 - E_{10}$$

in terms of the rest mass energies E_{10} and E_{20} and the total rest mass energy E_0 of the particles resulting from the interaction. To do so we use the conservation of total energy and linear momentum. We then eliminate the linear momentum by making use of the "useful triangle" relation.

The conservation of linear momentum and energy are expressed by

	BEFORE		DURING
(1)	p_1	=	p
(2)	E_1+E_{20}	=	E

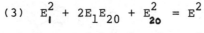

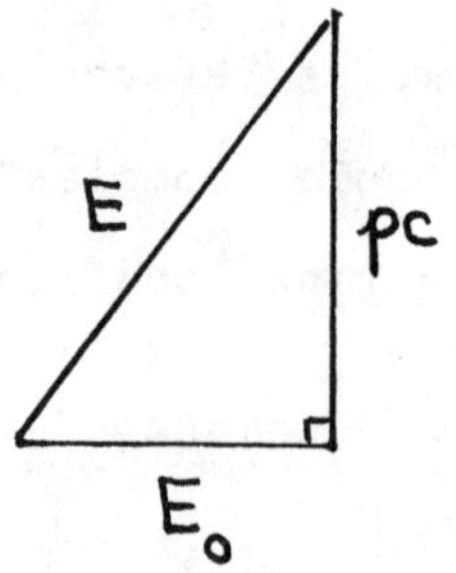

Useful Triangle

Square both sides of (2) to get

(3) $E_1^2 + 2E_1E_{20} + E_{20}^2 = E^2$

Next we use the useful triangle relation for E_1^2 and E^2,

$$E_1^2 = E_{10}^2 + (p_1c)^2; \quad E^2 = E_0^2 + (pc)^2$$

Substituting these expressions into (3) gives

$$E_{10}^2 + (p_1c)^2 + 2E_1E_{20} + E_{20}^2 = E_0^2 + (pc)^2$$

Because of linear momentum conservation ($p_1 = p$) we can cancel the terms $(p_1c)^2$ and $(pc)^2$. Next, we add and subtract $2E_{10}E_{20}$ on the left side to get

$$E_{10}^2 + 2E_{10}E_{20} + E_{20}^2 + 2(E_1 - E_{10})E_{20} = E_0^2$$

The first three terms equal $(E_{10} + E_{20})^2$. Solving for $K = E_1 - E_{10}$

(4) $K = [E_0^2 - (E_{10} + E_{20})^2]/2E_{20}$

This is the desired result. It expresses the minimum kinetic energy which the incident particle must have in order to produce a group of particles whose total rest mass energy is E_0.

If we consider interactions in which particles are created <u>without destroying the colliding particles</u> we can write

$$E_0 = E_{10} + E_{20} + D$$

where D is the total rest mass energy of the newly created particles. In terms of D

(5) $K = [D^2 + 2D(E_{10} + E_{20})]/2E_{20}$

For the collision of two protons in which a pion pair is created

$$E_{10} = E_{20} = 938.25 \text{ MeV}; \quad D = 2E_\pi = 2(139.58 \text{ MeV}) = 279.16 \text{ MeV}$$

For K, Eq 5 gives the same result obtained in Solution I,

$$K = (279.16)^2/2(938.25) + 2(279.16) \text{ MeV} = 600 \text{ MeV}$$

Equations (4) and (5) permit solutions for a wide variety of particle creation interactions. For example, the production of an electron-positron pair when a gamma ray photon collides with a massive nucleus can be treated by setting $E_{10} = 0$ (photon rest mass is zero).

One final observation: E_0 is the rest energy of the complex particle. It is also the rest energy of the system in the CM frame. At threshhold the product particles have zero kinetic energy in the CM frame. This means that we can interpret

$$D = E_o - (E_{10} + E_{20})$$

as the kinetic energy that is available to create new particles, as reckoned from the CM frame.

20.30 HINT: Make use of the solution developed for 20.29. Note especially the interpretation of D as kinetic energy available in the CM frame.

20.33 HINT: a) The conservation of total energy can be expressed as

$$E_{\Lambda} = E_p + E_{\pi}$$

$$7.993 m_{\pi} c^2 = 6.722 m_{\pi} c^2/(1-\beta_p^2)^{\frac{1}{2}} + m_{\pi} c^2/(1 - \beta_{\pi}^2)^{\frac{1}{2}}$$

Setting $\beta_p = 0$ allows you to solve for β_{π}.

b) The conservation of linear momentum can be expressed as

$$\underset{\sim}{p}_{\pi} + \underset{\sim}{p}_p = 0 \Rightarrow m_{\pi}\beta_{\pi}/(1 - \beta_{\pi}^2)^{\frac{1}{2}} = 6.722 m_{\pi}\beta_p/(1 - \beta_p^2)^{\frac{1}{2}}$$

Using the value of β_{π} determined in a) allows you to solve for β_p. With values of β_{π} and β_p you can determine the kinetic energies of the pion and proton.

20.34 HINT: Make use of the "useful triangle" to express E_{π} and E_p in terms of their momenta and rest energies. If p_{π} is expressed in units of $m_{\pi}c$ you have the equation

$$[p_{\pi}^2 + 1]^{\frac{1}{2}} + [p_{\pi}^2 + (6.722)^2]^{\frac{1}{2}} = 7.993$$

This can be solved numerically with a hand-held calculator. Note also that you can solve algebraically for p_{π} by noting that the equation

$$[x^2 + 1]^{\frac{1}{2}} + [x^2 + A]^{\frac{1}{2}} = B$$

has the solution

$$x = \{[(B^2+A-1)/2B]^2 - A\}^{\frac{1}{2}}$$

21.5 HINT: Use Eqs 21.4 and 21.5 to eliminate t_C. Then set $t_F = T$.

21.12 a) Heat capacity = mC = (3 kg)(0.900 kJ/kg·C°) = 2.70 kJ/C°

b) $Q = mC\Delta T$ = (2.70 kJ/C°)(10 C°) = 27.0 kJ

21.14 HINT: 1 gallon of water occupies a volume of 3.785×10^{-3} m^3.

21.15 HINT: Evaluate $Q = m\int_{T_i}^{T_f} C\,dT$

21.18 The mechanical energy acquired by a mass m falling a distance h with a constant acceleration g is mgh. The heat absorbed by a mass m of a substance with a specific heat C when its temperature rises by ΔT is $mC\Delta T$. Mayer's statement means that

$$mgh = mC\Delta T \qquad \text{when } h = 365 \text{ m}, \quad T = 1\ \text{C}^\circ,\ C = 1\ \text{kcal/kg·C}^\circ$$

With m = 1 kg,

$$mgh = (1\ \text{kg})(9.80\ \text{m/s}^2)(365\ \text{m}) = 3.58\ \text{kJ}$$

$$mC\Delta T = (1\ \text{kg})(1\ \text{kcal/kg·C}^\circ)(1\ \text{C}^\circ) = 1\ \text{kcal}$$

For these data 3.58 kJ = 1 kcal, so J = 3.58 kJ/kcal is the mechanical equivalent of heat.

21.22 HINT: Work done at constant volume is given by $P\Delta V$ where ΔV is the change in volume of a cube whose edge lengths increase from 1 m to 1.001 m.

21.24 The work equals the "area" under the graph of P vs V.

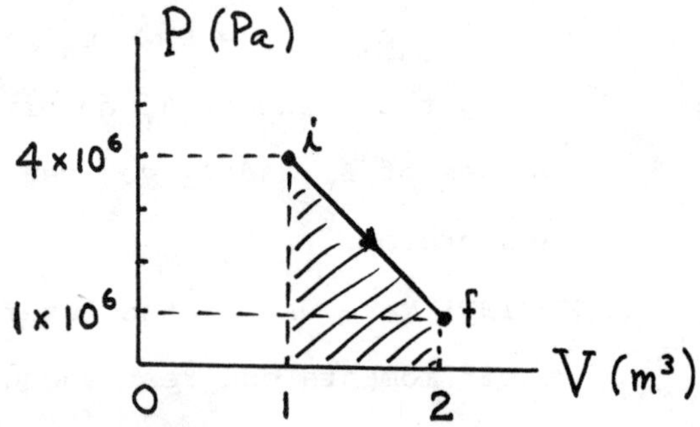

$$W = \text{"area"} = (\text{base})(\text{average height})$$

$$= (1\ \text{m}^3)(\tfrac{1}{2}[4 + 1] \times 10^6\ \text{Pa})$$

$$= 2.5 \times 10^6\ \text{J} = 2.5\ \text{MJ}$$

Note: 1 Pa m^3 = (1 N/m^2) m^3 = 1 N·m = 1 J

21.28 HINT: a) PV = constant = (Initial Pressure)(Initial Volume) = P_iV_i

c) $W = \int_{V_i}^{V_f} P\,dV = P_iV_i \int_{V_i}^{V_f} (dV/V)$ because $P = P_iV_i/V$

21.30 HINT: Recall $\Delta U = 0$ for a cyclic process. The first law for a cyclic process is expressed by Eq 21.20.

21.34 HINT: The area of the ellipse shown is πab.

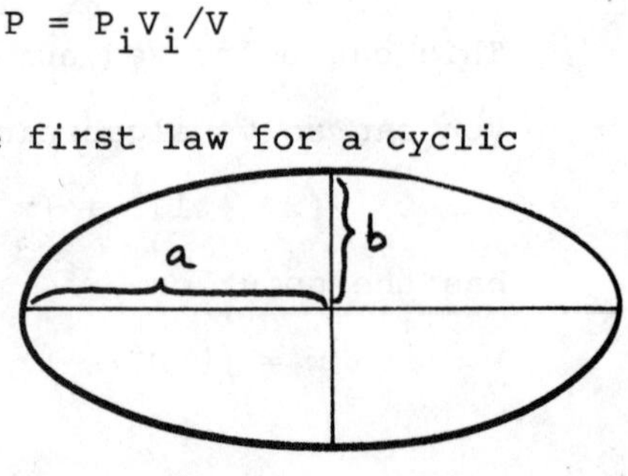

22.2 HINT: At $25^{\circ}C = 77^{\circ}F$ the tape would give a correct reading. At $85^{\circ}F$ the tape has expanded so the actual distances are larger than indicated by the tape. [For example, imagine using the tape to measure the distance between 2 points 1 meter apart. At $77^{\circ}F$ the tape would indicate a separation of 1 m. As the temperature increases two points on the tape that were originally 1 m apart would become more than 1 m apart. Such a tape would indicate a separation of less than 1 m for the two points whose actual separation is 1 m.]

The relation $\Delta L = \alpha L \Delta T$ lets you relate the actual distance to the apparent difference caused by thermal expansion.

22.4 The gap width increases. To see why, imagine a complete circular casting, with marks painted on it where cuts are to be made to produce the gap. If the solid casting is heated it expands, but remains circular. All points on the ring move away from one another. In particular, the lines marking the gap move farther apart - the gap widens when the casting is heated.

The width of the gap equals the width of the iron that was removed to create the gap.

$$L = L_o[1 + \alpha \Delta T] = 1.600[1 + (1.17 \times 10^{-5}/C^{o})(160\ C^{o})]\ cm$$
$$= 1.603\ cm$$

22.7 HINT: The period P is proportional to the square root of the length of the pendulum, $L = L_o(1 + \alpha \Delta T)$.

$$P = const \times L^{\frac{1}{2}} = const \times [1 + \alpha \Delta T]^{\frac{1}{2}}$$

If $P = 1$ s when $\Delta T = 0$, then in general

$$P = [1 + \alpha \Delta T]^{\frac{1}{2}}\ s$$

22.11 a) $V = RT/P = (8.314)(293)/[2(1.013 \times 10^{5}\ Pa)]\ m^3 = 0.012\ m^3$

b) For an isothermal expansion PV = RT = constant. If pressure is halved, volume doubles; $V_f = 0.024\ m^3$

c) For an adiabatic process PV^{γ} = constant = $P_iV_i^{\gamma}$, with $\gamma = 5/3$ for He.

$$V_f/V_i = [P_i/P_f]^{1/\gamma} = 2^{3/5} = 1.516$$
$$V_f = (1.516)(0.012\ m^3) = 0.0182\ m^3$$

d) $T_f = P_fV_f/R = (1.013 \times 10^{5}\ Pa)(0.0182)/8.314 = 222\ K$

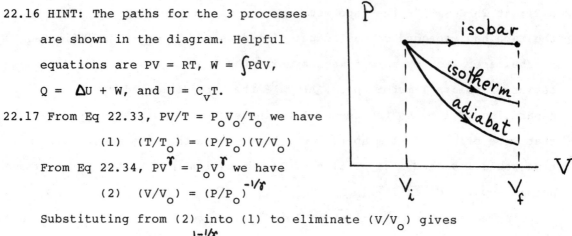

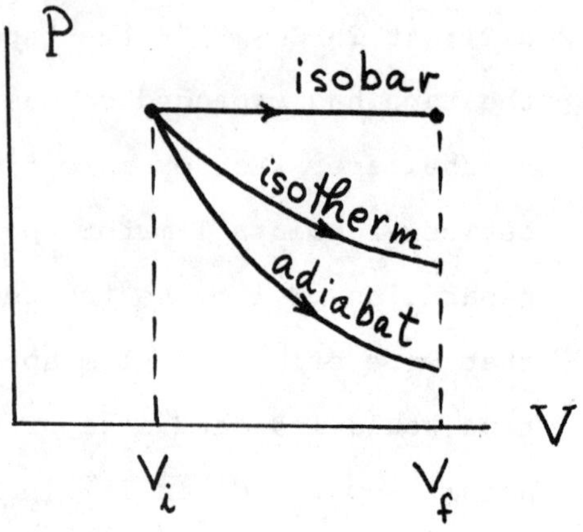

22.16 HINT: The paths for the 3 processes are shown in the diagram. Helpful equations are $PV = RT$, $W = \int PdV$, $Q = \Delta U + W$, and $U = C_V T$.

22.17 From Eq 22.33, $PV/T = P_oV_o/T_o$ we have

(1) $(T/T_o) = (P/P_o)(V/V_o)$

From Eq 22.34, $PV^{\gamma} = P_oV_o^{\gamma}$ we have

(2) $(V/V_o) = (P/P_o)^{-1/\gamma}$

Substituting from (2) into (1) to eliminate (V/V_o) gives

$$T/T_o = (P/P_o)^{1-1/\gamma}$$

22.23 The heat required to melt 100 grams of ice is

$$mL_f = (0.1\ \text{kg})(333.6\ \text{kJ/kg}) = 33.4\ \text{kJ}$$

This heat is supplied over a period of t = 31 min = 1860 s. The thermal power input to the ice is

$$P_{thermal} = mL_f/t = 33.4\ \text{kJ}/1860\ \text{s} = 17.9\ \text{W}$$

23.5 HINT: Energy conservation requires that the heat flux for the Al plate equal the heat flux for the Fe plate. It follows from Fourier's law that the 3:1 ratio of conductivities requires a 1:3 ratio of temperature drops.

23.8 HINT: Equation 23.8 is appropriate for heat flow in two dimensions.

23.9 The heat loss per meter is given by

$$q/L = -2\pi k \Delta T/\ln(r/r_i)$$

From Problem 23.8, q/L = 100 W/m when r_i = 1 cm and r = 2 cm.

a) With insulation 2 cm thick, r = 3 cm, r_i = 1 cm, and

$$q/L = (100\ \text{W/m})(\ln 2/\ln 3) = 63.1\ \text{W/m}$$

b) With insulation 3 cm thick, r = 4 cm, r_i = 1 cm, and

$$q/L = (100\ \text{W/m})(\ln 2/\ln 4) = 50\ \text{W/m}$$

c) With insulation 4 cm thick, r = 5 cm, r_i = 1 cm, and

$$q/L = (100\ \text{W/m})(\ln 2/\ln 5) = 43.1\ \text{W/m}$$

23.12 a) HINT: Separate the variables to get

$$\int_{T_1}^{T_2} dT = -(q/4\pi k)\int_{r_1}^{r_2} dr/r^2$$

23.17 For a steady state situation (constant T) the ceramic surface must radiate energy at the same rate electric energy is dissipated by the wire. The coating on the ceramic surface determines what temperature the surface must reach in order to radiate at the same rate it receives energy from the wire. From Eq 23.21

$$F_{net} = \epsilon\sigma(T^4 - T_e^4)$$

it follows that in order to radiate at equal rates the silver surface (ϵ = 0.3) must have a higher temperature than the black surface (ϵ = 0.9). Equating F_{net} for the silver and black surfaces (with T_e = 310 K, and T_{black} = 360 K) gives

$$0.3\not{\sigma}[T_{silver}^4 - (310)^4] = 0.9\not{\sigma}[(360)^4 - (310)^4]$$

Solving for T_{silver} gives

$$T_{silver} = [3(360)^4 - 2(310)^4]^{1/4} = 423 \text{ K}$$

23.19 The radiated power is

$$q = FA = \sigma A(T^4 - T_e^4); \qquad A = \text{radiating area}$$

With T_e = 273 + 27 = 300 K, and T = 273 + 32 = 305 K, A = 0.84 m^2,

$$q = 5.67 \times 10^{-8}(0.84)[(305)^4 - (300)^4] = 26.4 \text{ W}$$

The energy radiated in time t is Q = qt. With t = 1 day = 86,400 s,

$$Q = qt = (26.4 \text{ W})(86,400 \text{ s}) = 2.28 \times 10^6 \text{ J}$$

We can convert to kcal by using the conversion factor 1 kcal = 4.19 x 10^3 J

$$Q = 2.28 \times 10^6 \text{ J } (1 \text{ kcal}/4.19 \times 10^3 \text{ J}) = 544 \text{ kcal}$$

23.26 HINT: Equation 23.27 relates the surface temperatures of Earth and the sun. Write down by analogy the corresponding equation which relates the surface temperatures of Pluto and the sun. Then form a ratio so as to eliminate the surface temperature of the sun. Take T_E = 285 K as in Example 7.

23.27 HINT: The <u>net</u> outward rate of energy flow between the surfaces is

$$q_{net} = \underbrace{4\pi R^2 T^4}_{\substack{\text{outward from}\\ \text{inner sphere}}} - \underbrace{4\pi(2R)^2 T_o}_{\substack{\text{inward from}\\ \text{outer surface}}}$$

23.28 a) HINT: Equating the rate of radiation with the rate of heat loss gives

$$-mC(dT/dt) = \epsilon\sigma AT^4$$

Separate variables to get

$$-\int_{T_0}^{T}(dT/T^4) = (\epsilon\sigma A/mC)\int_0^t dt$$

24.4 The work output is

$$W = Q_{net} = 4Q_o - 3Q_o = Q_o$$

The heat absorbed is $4Q_o$. The efficiency is

$$\eta = W/Q_{in} = Q_o/4Q_o = 1/4$$

24.10 The coefficient of performance is defined by

$$COP = Q_L/W \quad \Rightarrow \quad 1 + COP = (W + Q_L)/W$$

The first law shows that $W + Q_L = Q_H$, so

$$1 + COP = Q_H/W$$

The efficiency is defined by $\eta = W/Q_H$. Hence

$$\eta = 1/(1 + COP)$$

24.12 HINT: a) Use the result of Problem 24.10. b) Use the definition of the COP.

24.14 HINT: Remember that the internal energy of an ideal gas depends on temperature only.

24.15 HINT: Along the 400 K isotherm, $\Delta U = 0$, so the first law gives $Q_{abs} = W$. As the gas expands along the 400 K isotherm, $PV = P_1V_1$, so $P = P_1V_1/V$. The work performed along the isotherm is

$$W = \int_1^2 PdV = P_1V_1 \int_1^2 dV/V = P_1V_1 \ln 2$$

24.17 The Carnot efficiency is given by $\eta = 1 - T_L/T_H$. If T_L is reduced η increases. If T_H is increased η increases. You can choose a pair of values for T_L and T_H and show numerically that a 25 K decrease in T_L results in a larger increase in efficency thatn does a 25 K increase in T_H.

A general proof showing that a reduction in T_L produces a greater increase in η than does an equal increase in T_H goes like this: Let $x > 0$ denote the temperature decrease in T_L. Then

$$\eta_- = 1 - (T_L - x)/T_H$$

is the resulting efficiency. The efficiency resulting from an equal increase in T_H is

$$\eta_+ = 1 - T_L/(T_H + x)$$

We can prove that $\eta_- > \eta_+$ by showing that $\eta_- - \eta_+ > 0$.

$$\eta_- - \eta_+ = 1 - (T_L - x)/T_H - [\,1 - T_L/(T_H + x)]$$
$$= [T_L T_H - (T_L - x)(T_H + x)]/[T_H(T_H + x)]$$
$$= [x(T_H + x - T_L)]/[T_H(T_H + x)]$$

All factors are positive [including $(T_H + x - T_L)$ because $T_H > T_L$], so

$$\eta_- - \eta_+ > 0,$$

which shows that a decrease in T_L produces a larger increase in efficiency than does an equal increase in T_H.

24.26 a) $\Delta S = \int_i^f dS = \int_i^f dQ/T = mC\int_{T_i}^{T_f} dT/T = mC\cdot\ln(T_f/T_i)$

$= (0.020 \text{ kg})(0.39 \text{ kJ/kg}\cdot C^o)\ln(293/273) = 0.551 \text{ J/C}^o = 0.551 \text{ J/K}$

The change of units in the final step makes use of the fact that a temperature <u>difference</u> of 1 celsius degree (1 C^o) equals a temperature difference of 1 kelvin.

b) $Q = mC\Delta T = (0.020 \text{ kg})(0.39 \text{ kJ/kg}\cdot C^o)(20\ C^o) = 156 \text{ J}$

c) If the 156 J were absorbed at the average temperature of 283 K the entropy change would be

$$\Delta S = Q/T = 156 \text{ J}/283 \text{ K} = 0.551 \text{ J/K}$$

This is the same result obtained in part a). <u>In general</u> the results obtained using the methods of a) and c) differ. For relatively small temperature changes (as in this problem) the <u>numerical</u> results are equal. You might try repeating the calculations for temperatures of 0^oC and 200^oC to verify that the two methods do not give identical results.

24.30 HINT: For a reversible process, TdS = dQ, the heat exchange.

24.31 HINT: Equation 24.28 gives the entropy change for an ideal gas.

25.7 The relation $n\sigma\lambda = 1$ can be used to determine the mean free path provided n is the number of cars per unit <u>area</u> and σ is the effective <u>width</u> of a car. With $n = 22/(300 \text{ m}^2)$ and $\sigma = 2$ m,

$$\lambda = 1/n\sigma = (300/44) \text{ m} = 6.8 \text{ m}$$

Thus, the mean free path of a car is approximately 7 m.

25.8 HINT: See the solution for problem 25.7. Another approach is to treat the balls as a gas which occupies a <u>volume</u> of 4 ft x 8 ft x 2.25 in.

25.12 HINT: Show that

$$\langle v^2 \rangle - \langle v \rangle^2 = (1/4)(v_1 - v_2)^2 > 0$$

which proves $\sqrt{\langle v^2 \rangle} > \langle v \rangle$

25.13 HINT: There is a <u>distribution</u> of molecular speeds. A small but significant fraction of the atoms are moving at speeds in excess of 5 times the rms speed. See Section 25.4.

25.16 $(3/2)kT = 3 \times 10^{-19}$ J

$$T = 3 \times 10^{-19} \text{ J}/(1.5)(1.38 \times 10^{-23} \text{ J/K}) = 14{,}500 \text{ K}$$

This result shows that the kinetic energy of conduction electrons <u>does not</u> stem from <u>thermal</u> motions. It is a quantum mechanical effect - a manifestation of the quantum <u>zero point</u> kinetic energy of the electrons.

25.18 HINT: Imagine a volume with a cross-sectional area $A = 1$ cm^2 and a length $L = vt$ where v is a "typical" molecular speed and $t = 1$ s. Approximately half of the molecules in this volume will cross the area in 1 s. Now, imagine that the area A is 1 cm^2 of your forehead.

25.26 HINT: From Example 7, the most probable speed is 390 m/s. The number of molecules with speeds greater than 780 m/s corresponds to the area under the f(v) graph for $v \geq 780$ m/s. You can approximate this area by the area of a triangle.

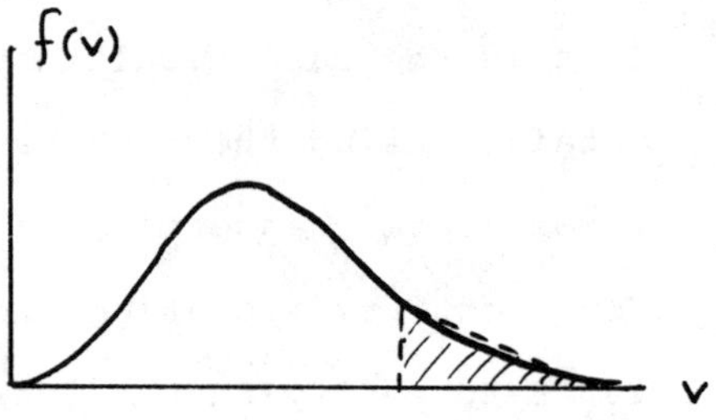

26.3 The number of electrons (N) times the elctron charge ($e = 1.60 \times 10^{-19}$ C) equals the total charge Q

$$Q = Ne$$

If we set $Q = 10^{-10}$ C then N is the number of electrons that can be present at the limit of detectability

$$N = Q/e = 10^{-10}\ \text{C}/1.60 \times 10^{-19}\ \text{C} \simeq 6 \times 10^{8}$$

26.7 HINT: The initial force is attractive so the charges must have opposite signs. Let $-q_1$ and q_2 denote the charges on the two spheres. The initial force is

$$F_i = k_e q_1 q_2 / r^2$$

When contact is established the spheres share the net charge so that each has a charge of $\frac{1}{2}(q_2 - q_1)$. The final (repulsive) force is

$$F_f = k_e[\tfrac{1}{2}(q_2 - q_1)]^2/r^2$$

These two equations can be solved for the product $q_1 q_2$ and the quantity $(q_2 - q_1)^2$. Knowing $q_1 q_2$ and $(q_2 - q_1)^2$ you can determine q_1 and q_2.

26.13 a) $F = k_e e^2/r^2 = 8.99 \times 10^9 (1.60 \times 10^{-19})^2/(0.529 \times 10^{-10})^2$

$$= 8.22 \times 10^{-8}\ \text{N}$$

b) $a = F/m = v^2/r \Rightarrow v = [Fr/m]^{\frac{1}{2}}$

$$v = [(8.22 \times 10^{-8})(0.529 \times 10^{-10})/(0.911 \times 10^{-30})]^{\frac{1}{2}}$$

$$= 2.19 \times 10^{6}\ \text{m/s}$$

26.22 HINT: The net force on each sphere is zero. Choose one sphere as a body, and resolve the forces acting on it into horizontal and vertical components. The resulting pair of equations can be solved for the charge.

26.23 HINT: Doubling the separation reduces the electric force by a factor of four. The condition for equilibrium in the horizontal direction involves $\sin\theta$ and the electric force. For small angles (as here) $\sin\theta \simeq \theta$.

26.24 HINT: Let $T = Q+q$ denote the total charge. T is a constant. Express the force F in terms of T and Q. The charge distribution that gives the maximum force can be determined by imposing the condition

$$dF/dQ = 0$$

26.27 The forces exerted by charges 4 and 2 cancel each other so the net force is that exerted by charge 1. From the geometry of Figure 2 the distance between charges 1 and 3 is $r = 1/\sqrt{2}$ m. The magnitude of the net force is

$$F = k_e Q^2/r^2 = (8.99 \times 10^9)(10^{-6})/(1/2) = 0.018 \text{ N}$$

This force is directed at 45° above the positive x-axis.

26.30 HINT: Q cannot be positive if all 3 charges are to be in equilibrium. Further, Q cannot be in equilibrium if it lies to the left of the 1 μC charge or to the right of the 3 μC charge. One method of solving for the position of Q leads to a quadratic equation. Only one of the two mathematical solutions corresponds to an equilibrium position.

26.34 HINT: Exploit symmetry!

26.35 With q_o on the x-axis, the attractive force extented by -q at the origin must be balanced by the x-components of the 2 positive charges. From the figure,

$$F = k_e q q_o/(s + \tfrac{1}{2}a\sqrt{3})^2 + k_e q q_o(2\cos\theta)/r^2 = 0$$

With $r = \tfrac{1}{2}a/\sin\theta$; $s = \tfrac{1}{2}a/\tan\theta$

we get

$$(4k_e q q_o/a^2)[2\cos\theta \sin^2\theta - 1/(\sqrt{3} + \cot\theta)^2] = 0$$

The angle θ is determined by requiring that the factor in [ ] equal zero, or

$$2\cos\theta \sin^2\theta(\sqrt{3} + \cot\theta)^2 = 1$$

With $\theta = 90^\circ$ the left side is zero. With $\theta = 80^\circ$ the left side is 1.226, so the correct value of θ lies between 80° and 90°. You can use your calculator to show that $\theta = 81.7^\circ$. The equilibrium position is

$$x = \tfrac{1}{2}a(\sqrt{3} + \cot 81.7^\circ) = 0.939a$$

26.37 HINT: The force is given by

$$F = 2k_e q q_o\{(1+x)/[1+(1+x)^2]^{3/2} - (1-x)/[1+(1-x)^2]^{3/2}\}$$

Note that for $x \ll 1$ the terms $[1 + (1+x)^2]^{3/2}$ can be approximated as

$$[1 + 1 \pm 2x + x^2]^{3/2} \cong [2 \pm 2x]^{3/2} = (2)^{3/2}[1 \pm x]^{3/2}$$

Using the binomial expansions of $(1 \pm x)^{-\frac{1}{2}}$ lets you show

$$F = -(k_e q q_o/\sqrt{2})x$$

Recall that $F = -kx$ leads to simple harmonic motion.

26.38 By symmetry the net force acts along the +y-axis. The force exerted on q_o by an infinitesimal element of charge dq is

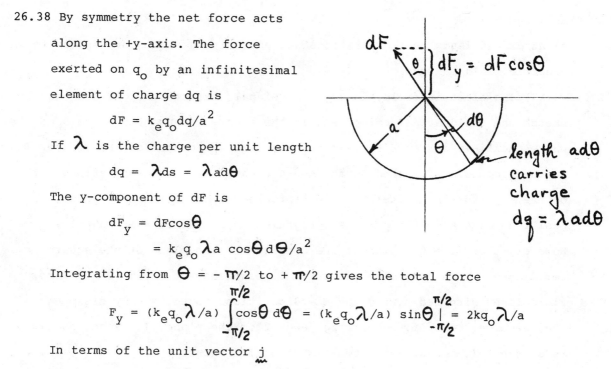

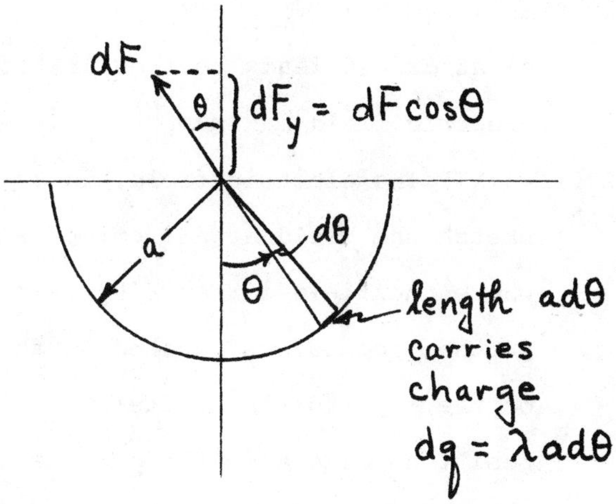

$$dF = k_e q_o dq/a^2$$

If λ is the charge per unit length

$$dq = \lambda ds = \lambda a d\theta$$

The y-component of dF is

$$dF_y = dF\cos\theta$$
$$= k_e q_o \lambda a \cos\theta \, d\theta/a^2$$

Integrating from $\theta = -\pi/2$ to $+\pi/2$ gives the total force

$$F_y = (k_e q_o \lambda/a)\int_{-\pi/2}^{\pi/2}\cos\theta \, d\theta = (k_e q_o \lambda/a)\sin\theta \Big|_{-\pi/2}^{\pi/2} = 2kq_o\lambda/a$$

In terms of the unit vector $\underset{\sim}{j}$

$$\underset{\sim}{F} = (2k_e q_o \lambda/a)\underset{\sim}{j}$$

27.6 The gravitational field equals the gravitational force per unit mass, which is simply the acceleration due to gravity.

$$a = F/m = (GM_E m/r^2)/m = GM_E/r^2$$

$$G = 6.67 \times 10^{-11}\ N{\bullet}m^2/kg^2 \ ;\ M_E = 5.98 \times 10^{24}\ kg,$$

a) at the surface of the earth,

$r = 6.37 \times 10^6$ m (polar radius), or

$r = 6.38 \times 10^6$ m (equatorial radius)

The gravitational field obtained using the equatorial radius is

$$a = (6.67 \times 10^{-11})(5.98 \times 10^{24})/(6.38 \times 10^6)^2 = 9.80\ N/kg$$

b) At the lunar surface ($r = 3.84 \times 10^8$ m) the earth's gravitational field is

$$a = (6.67 \times 10^{-11})(5.98 \times 10^{24})/(3.84 \times 10^8)^2 = 0.00270\ N/kg$$

NOTE: The gravitational field at the lunar surface is much larger than 0.00270 N/kg. The moon's mass is the primary source of the field at the lunar surface.

27.9 HINT: a) $\mathbf{F} = q_o\mathbf{E}$

b) $dE/dx = 0$ leads to a quadratic equation. Only one root corresponds to an electric field maximum.

27.11 HINT: The electric field has a 3-dimensional character. It is not easy to sketch the field at all points within the square because there are 5 points where the field is zero. By symmetry E = 0 at the center of the square. The 4 other points are located on the + x-axes and + y-axes at positions 0.773 of the way from the center-to-edge. (Problem 26.36 establishes these positions. See also the computer programs at the end of this manual.) However, these E = 0 positions are confined to the plane of the square. Immediately above and below these E = 0 positions the field is not zero. Field lines directed toward the E = 0 positions do not simply disappear - they diverge. Some of them diverge away from the plane. It is not possible to sketch this feature on a two-dimensional diagram.

27.15 HINT: Method 1. Note that $\mathbf{E}$ can be expressed as

$$\mathbf{E} = \mathbf{k}(q/4\pi\epsilon_o)\{(1 - a/z)^{-2} - (1 + a/z)^{-2}\}$$

Then use the binomial expansions of $(1 \pm a/z)^{-2}$. You must retain the first 4 terms in each expansion to get the first 2 non-zero terms in $\mathbf{E}$.

Method 2. Start with the above expression for $\mathbf{E}$, and note that

$$\{(1 - a/z)^{-2} - (1 + a/z)^{-2}\} = (2a/z)(1 - a^2/z^2)^{-2}$$

Then use a 2-term binomial expansion for $(1 - a^2/z^2)^{-2}$

27.16 a) The total electric field is the vector sum of the fields of the two point charges. The field of +q is directed away from that charge.

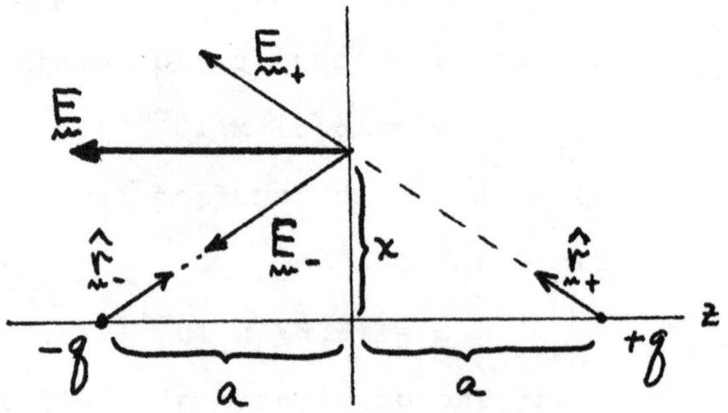

$$\mathbf{E}_+ = \hat{\mathbf{r}}_+(q/4\pi\epsilon_o)[1/(a^2 + x^2)]$$

where $\hat{\mathbf{r}}_+$ is a unit vector.

The field of -q is directed toward that charge.

$$\mathbf{E}_- = -\hat{\mathbf{r}}_-(q/4\pi\epsilon_o)[1/(a^2 + x^2)]$$

where $\hat{\mathbf{r}}_-$ is a unit vector. The field $\mathbf{E}(x,0,0)$ is the vector sum of $\mathbf{E}_+$ and $\mathbf{E}_-$.

$$\mathbf{E} = \mathbf{E}_+ + \mathbf{E}_- = (q/4\pi\epsilon_o)(\hat{\mathbf{r}}_+ - \hat{\mathbf{r}}_-)/(a^2 + x^2)$$

27.16 b) From the figure we can see that the vector $\hat{\mathbf{r}}_+ - \hat{\mathbf{r}}_-$ is directed along the negative z-axis and has a magnitude $2\cos\theta$. In terms of the unit vector $\mathbf{k}$,

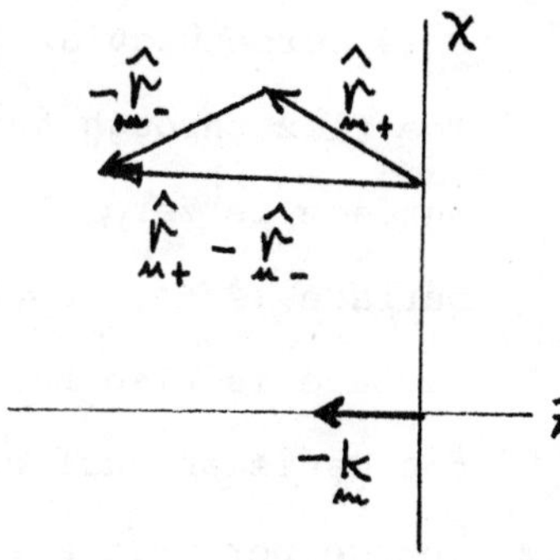

$$\hat{\mathbf{r}}_+ - \hat{\mathbf{r}}_- = -2\mathbf{k}\cos\theta$$
$$= -2\mathbf{k}a/[a^2 + x^2]^{\frac{1}{2}}$$

c) Using the result of b) in a) establishes

$$\mathbf{E}(x,0,0) = -\mathbf{k}[2aq/4\pi\epsilon_0]\{1/(x^2 + a^2)^{3/2}\}$$

d) For $x >> a$, $(x^2 + a^2)^{3/2} \simeq x^3$, and

$$\mathbf{E}(x,0,0) = -\mathbf{k}[2aq/4\pi\epsilon_0]/x^3$$

The x^{-3} behavior is typical of a dipole. The quantity 2aq is the electric dipole moment.

27.19 HINT: Show that the electric field is

$$E_x = (q/4\pi\epsilon_0)\{1/(x-a)^2 - 2/x^2 + 1/(x+a)^2\}$$

Then multiply by appropriate factors to obtain the common denominator

$$x^2(x - a)^2(x + a)^2 = x^2(x^2 - a^2)^2$$

The x^2 term in the denominator is canceled by an identical factor in the numerator. Note that for $x >> a$, $(x^2 - a^2)^2 \simeq x^4$.

27.25 HINT: Because $\mathbf{E}$ is parallel to the x-axis, $\mathbf{E}\cdot d\mathbf{A}$ is zero on four of the six faces. Only the two faces perpendicular to the x-axis contribute to the net flux. If the electric field were simply $\mathbf{i}E_0$ (E_0 = constant) what would be the flux?

27.27 HINT: $\int_0^\infty y\,dy/(y^2 + a^2)^{3/2} = -(y^2 + a^2)^{-\frac{1}{2}}\Big|_0^\infty = 1/a$

27.35 The figures show top and side views of the appropriate Gaussian surface. By symmetry $\underset{\sim}{E}$ is perpendicular to the sheet of charge. The flux through the sides of the Gaussian surface is zero. The flux through the top surface is EA. The flux through the bottom surface is also EA. The charge enclosed by the Gaussian surface is σA, where σ is the charge per unit area. Gauss' law becomes

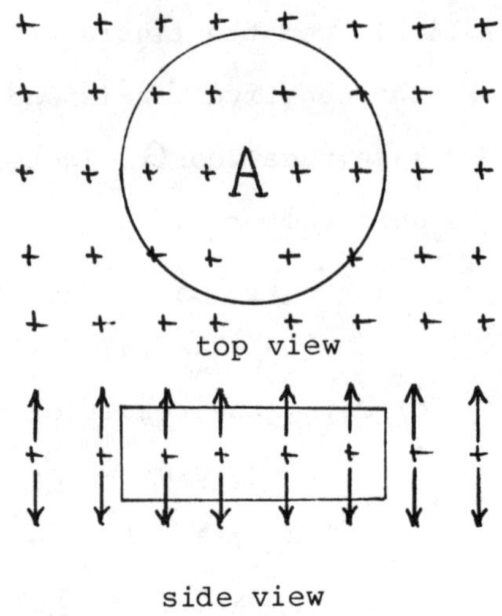

$$2EA = \sigma A/\epsilon_o$$

which gives

$$E = \sigma/2\epsilon_o$$

27.36 HINT: Use the result of Problem 27.35 ($E = \sigma/2\epsilon_o$). Superimpose the fields of the two charge sheets - noting that their directions are related to the signs of the charges.

27.39 HINT: Use a sphere of radius $r \leqslant R$ for your Gaussian surface. By symmetry $\underset{\sim}{E}$ is normal to the Gaussian surface and has the same magnitude at all points on that surface. For $r \leqslant R$ the charge enclosed by a sphere of radius r is proportional to the volume of that sphere

$$q_{enclosed} = \text{(charge per unit volume)(volume of Gaussian sphere)}$$
$$= (Q/4\pi R^3/3)(4\pi r^3/3) = Qr^3/R^3$$

27.42 By symmetry the field is radial with a magnitude that depends only on the distance from the center of the sphere (r). The flux through the Gaussian surface, a sphere of radius r, is $\Phi = E\cdot 4\pi r^2$. Gauss' law gives

$$\Phi = E\cdot 4\pi r^2 = q_{enclosed}/\epsilon_o, \text{ and so } E = q_{enclosed}/4\pi\epsilon_o r^2$$

For $r < 0.1$ m $q_{enclosed} = 0$ and so $E = 0$.

For $r > 0.1$ m $q_{enclosed} = 6.0 \times 10^{-10}$ C, and

$$E = (6 \times 10^{-10})(8.99 \times 10^9)/r^2 = 5.39/r^2 \text{ N/C}$$

a) For $r = 0.05$ m < 0.1 m, $E = 0$

b) For $r = 0.2$ m > 0.1 m, $E = 5.39/(0.2)^2 = 135$ N/C

27.47 HINT: Use a Gaussian surface embedded in the outer conductor, and recall that E = 0 <u>inside</u> a conductor in electrostatic equilibrium. This choice of Gaussian surface guarantees that E = 0 at all points on the Gaussian surface. What does Gauss' law tell you about the net charge enclosed by the surface?

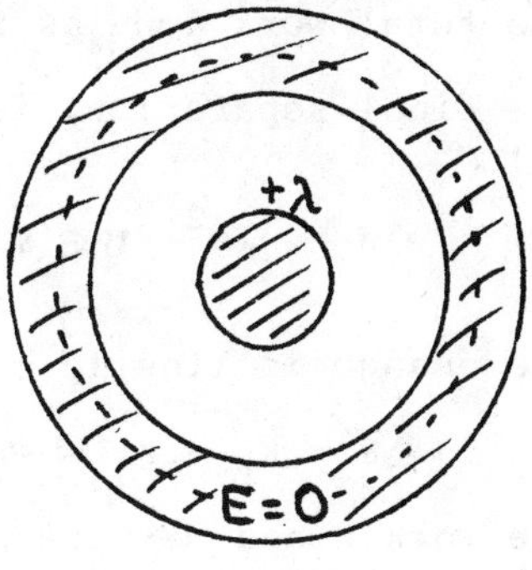

27.48 HINT: Exploit symmetry. The average field $\bar{E}$ is defined by the relation

$$\bar{E}A = \int \mathbf{E} \cdot d\mathbf{A}$$

where the integral is over the surface of area A.

27.50 a) The acceleration follows from Eq 27.19.

$$a = eE/m = (1.60 \times 10^{-19})(1.2 \times 10^{6})/(1.67 \times 10^{-27})$$
$$= 1.15 \times 10^{14} \text{ m/s}^2$$

b) With $v_o = 0$, $v^2 = v_o^2 + 2ax$ gives

$$v = [2ax]^{\frac{1}{2}} = [2(1.15 \times 10^{14})4]^{\frac{1}{2}} = 3.03 \times 10^{7} \text{ m/s}$$

c) The acceleration is constant so we can use $x = \frac{1}{2}at^2$ to find

$$t = [2x/a]^{\frac{1}{2}} = [2(4)/1.15 \times 10^{14}]^{\frac{1}{2}} = 2.64 \times 10^{-7} \text{ s}$$

27.54 HINT: Use the work-energy principle. The work done by the electric field is -eEx. The energy of the photon is transformed into work to free the electron and the kinetic energy of the liberated electron - there being an overall conservation of energy.

27.55 The attractive force between the electron (charge -e) and its image (charge +e) when they are separated by a distance x is

$$F = e^2/4\pi\epsilon_o x^2$$

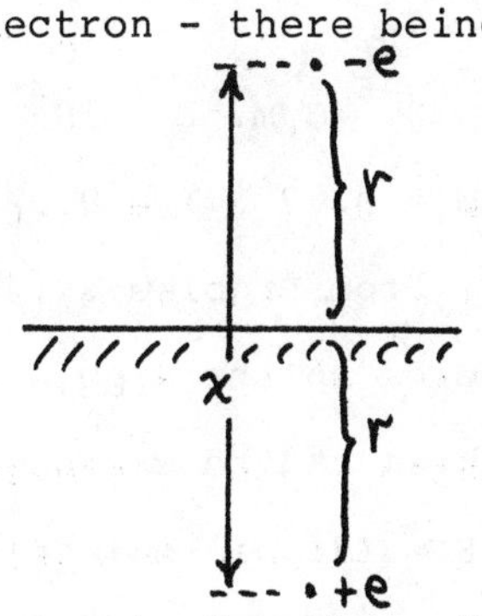

This attractive force performs a negative amount of work as the charge moves away from the surface (and thereby away from its image charge).

$$\mathbf{F} \cdot d\mathbf{x} = -Fdx = -e^2dx/4\pi\epsilon_o x^2$$

The total work done as x ranges from $2r_o$, the initial separation, to 2r, the final separation, is

$$W = -(e^2/4\pi\epsilon_o)\int_{2r_o}^{2r} dx/x^2 = -(e^2/4\pi\epsilon_o)[1/2r_o - 1/2r]$$

The change in kinetic energy of the electron is

$$\Delta K = K_f - K_i = -K_i$$

The work W and ΔK are related by the work-energy principle, $W = \Delta K$, which becomes

$$(-e^2/8\pi\epsilon_o)[1/r_o - 1/r] = -K_i$$

Solving for r gives

$$r = r_o/[1 - K_i r_o(8\pi\epsilon_o/e^2)]$$

With $r_o = 5 \times 10^{-11}$ m, $K_i = 1.60 \times 10^{-19}$ J, and $e^2/8\pi\epsilon_o = 1.15 \times 10^{-28}$ J•m,

$$r = 5 \times 10^{-11}\ \text{m}/[1 - (1.60 \times 10^{-19})(5 \times 10^{-11})/2.30 \times 10^{-28})]$$

$$= 5.37 \times 10^{-11}\ \text{m}$$

27.58 HINT: The rotational dynamics are governed by

$$I\alpha = \tau \quad ; \quad \alpha = d^2\theta/dt^2$$

The restoring torque is

$\tau = -pE\sin\theta$. For small angular displacements $\sin\theta \simeq \theta$.

27.59 The torque is $\tau = pE\sin\theta$. The work done by the torque is

$$W = \int_0^{\pi}\tau\, d\theta = pE\int_0^{\pi}\sin\theta\, d\theta = pE(-\cos\theta)\Big|_0^{\pi} = 2pE$$

With $p = 10^{-6}$ C m, $E = 10^4$ N/C,

$$W = 0.02\ \text{N•m} = 0.02\ \text{J}$$

27.62 HINT: a) From Problem 27.44, the electric field is given by $E = \lambda/2\pi\epsilon_o r$.

c) The force on the dipole is given by Eq 27.26, which in the notation of this problem can be expressed as

F = (dipole moment)(eletric field gradient) = p(dE/dr)

28.1 a), b), c) The electric field is constant so Eq 28.2 can be used.

$$W_{AB} = q\underset{\sim}{E} \cdot \Delta\underset{\sim}{s}$$

The net vector displacement $\Delta\underset{\sim}{s}$ is the same for all three paths and so W_{AB} is the same for all three paths. In terms of the unit vectors $\underset{\sim}{i}$ and $\underset{\sim}{j}$

$$\Delta\underset{\sim}{s} = 4\underset{\sim}{i} + 4\underset{\sim}{j} \text{ m}$$

With $q = 2 \times 10^{-7}$ C and $E = 10^4\underset{\sim}{i}$ N/C,

$$W_{AB} = (2 \times 10^{-7}\text{C})(10^4 \text{ N/C})\underset{\sim}{i}\cdot(4\underset{\sim}{i} + 4\underset{\sim}{j}) \text{ m}$$

Recalling that $\underset{\sim}{i}\cdot\underset{\sim}{i} = 1$, $\underset{\sim}{i}\cdot\underset{\sim}{j} = 0$ gives

$$W_{AB} = (2 \times 10^{-7})(10^4)(4) = 8 \times 10^{-3} \text{ N}\cdot\text{m} = 8 \times 10^{-3} \text{ J}$$

d) If B denotes the position (4,4) and A denotes the position (0,0) the potential difference $V_{(4,4)} - V_{(0,0)} = V_B - V_A$ follows from Eq 28.6

$$V_B - V_A = -W_{AB}/q = -(8 \times 10^{-3} \text{ J})/(2 \times 10^{-7} \text{ C}) = -40{,}000 \text{ V}$$

28.3 The special relativity expression for kinetic energy is given by Eq 20.18.

$$K_{SR} = mc^2\{[1 - v^2/c^2]^{-\frac{1}{2}} - 1\}$$

We use the binomial theorem (Appendix 4) to expand $[1 - v^2/c^2]^{-\frac{1}{2}}$ and retain only the first three terms. This gives

$$K_{SR} = mc^2\{[1 + \tfrac{1}{2}(v/c)^2 + (3/8)(v/c)^4] - 1\}$$
$$= \tfrac{1}{2}mv^2\{1 + (3/4)(v/c)^2\}$$

The Newtonian expression for kinetic energy is

$$K_{Newton} = \tfrac{1}{2}mv^2$$

The ratio we want to evaluate is

$$(K_{SR} - K_{Newton})/K_{SR} = 1 - K_{Newton}/K_{SR}$$
$$= 1 - \tfrac{1}{2}mv^2/\tfrac{1}{2}mv^2\{1 + (3/4)(v/c)^2\}$$
$$= 1 - \{1 + (3/4)(v/c)^2\}^{-1}$$

Consistent with our earlier use of the binomial expansion we can write

$$\{1 + (3/4)(v/c)^2\}^{-1} \simeq 1 - (3/4)(v/c)^2$$

Using this result gives the sought-for expression

$$(K_{SR} - K_{Newton})/K_{SR} = 1 - \{1 - (3/4)(v/c)^2\} = (3/4)(v/c)^2$$

28.4 HINT: b) A useful relation in special relativity is (See Problem 20.23)

$$v = pc^2/E$$

where p is the linear momentum, and E is the total energy. From

$$E^2 = E_o^2 + p^2c^2, \text{ and } E = E_o + K \quad (K = \text{kinetic energy})$$

we get

$$p^2c^2 = K^2 + 2KE_o + \cancel{E_o^2} - \cancel{E_o^2}$$

$$pc = [K^2 + 2KE_o]^{\frac{1}{2}},$$

from which it follows that

$$v/c = pc/E = [K^2 + 2KE_o]^{\frac{1}{2}}/(E_o + K)$$

28.10 $V_A - V_B = +12\text{ V} - 0 = 12\text{ V} \Rightarrow V_A = 12\text{ V}$

$V_A - V_C = +24\text{ V} \Rightarrow V_C = V_A - 24\text{ V} = 12\text{ V} - 24\text{ V} = -12\text{ V}$

28.11 HINT: E is related to V by Eq 28.17.

28.14 a) The kinetic energy equals the electric work done

$$K = q\Delta V = (1.60 \times 10^{-19}\text{ C})(14 \times 10^6\text{ V}) = 2.24 \times 10^{-12}\text{ J}$$

b) A proton accelerated through a potential difference of one volt increases its kinetic energy by one electron volt. For a 14 million volt potential difference the kinetic energy increase is 14 million electron volts, or 14 MeV.

c) The relation derived above in Problem 28.4 is useful.

$$v/c = [K^2 + 2KE_o]^{\frac{1}{2}}/(E_o + K)$$

With K = 14 MeV and E_o = 938 MeV

$$v/c = [14^2 + 2(14)(938)]^{\frac{1}{2}}/(938 + 14) = 0.171$$

The speed of the proton is

$$v = 0.171c = 0.171(3 \times 10^8\text{ m/s}) = 5.13 \times 10^7\text{ m/s}$$

If nonrelativistic mechanics is used to determine the speed, $K = \frac{1}{2}mv^2$, and

$$v = [2K/m]^{\frac{1}{2}} = [2(2.24 \times 10^{-12})/(1.67 \times 10^{-27})]^{\frac{1}{2}}$$

$$= 5.18 \times 10^7\text{ m/s}$$

28.19 HINT: Exploit symmetry.

28.22 HINT: The electric field cannot be zero between the two charges. You can convince yourself of this fact by imagining a test charge between the two charges. Both charges exert forces on the test charge in the same direction. An expression for E_x that is valid for positions that are <u>not</u> between the two charges is

$$E_x = (1/4\pi\epsilon_o)[q/x^2 - 2q/(2-x)^2]$$

The potential is given by

$$V = (1/4\pi\epsilon_o)[q/|x| - 2q/|2-x|]$$

28.25 The work to bring the charge q from infinity to a distance r from the charge Q is

$$W = \int_{\infty}^{r} \underset{\sim}{F}\cdot d\underset{\sim}{r}$$

where $\underset{\sim}{F}$ is the force exerted on q against the Coulomb force exerted by Q. For the purpose of indicating the direction of $\underset{\sim}{F}$ we suppose that Q and q have the same sign. Then the force $\underset{\sim}{F}$ which works to overcome the Coulomb force is directed from q toward Q.

$$\underset{\sim}{F}\cdot d\underset{\sim}{r} = -Fdr = -(qQ/4\pi\epsilon_o)dr/r^2$$

The subtle point in this problem is that $|dr| = -dr$, the reason being that dr is negative for a displacement in which r decreases. The work is

$$W = (Qq/4\pi\epsilon_o)\int_{\infty}^{r} -dr/r^2 = (Qq/4\pi\epsilon_o)(1/r)$$

28.26 HINT: At positions where the potential is a maximum, $dV/dy = 0$.

28.27 HINT: As in the gravitational analog of this problem, the total energy

E = kinetic energy + Coulomb potential energy

is conserved. The final configuration ($r_f \rightarrow \infty$, $v_f \rightarrow 0$) is one for which the total energy is zero.

28.33 The principle of superposition lets us express the potential set up by the disk as the sum (integral) of the potentials due to a set of concentric rings of charge. Let σ denote the charge per unit area on the surface of the disk. A ring of radius r and width dr has a surface area $2\pi r dr$ and carries a charge

$$dQ = 2\pi r dr\, \sigma$$

The potential set up by this ring of charge is

$$dV = dQ/(4\pi\epsilon_o)[x^2 + r^2]^{\frac{1}{2}}$$
$$= 2\pi\sigma\, rdr/(4\pi\epsilon_o)[x^2 + r^2]^{\frac{1}{2}}$$

Integrating from $r = 0$ to $r = a_o$ sums the potentials for the set of rings that collectively make up the disk.

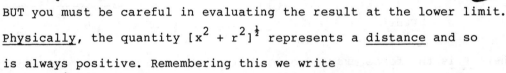

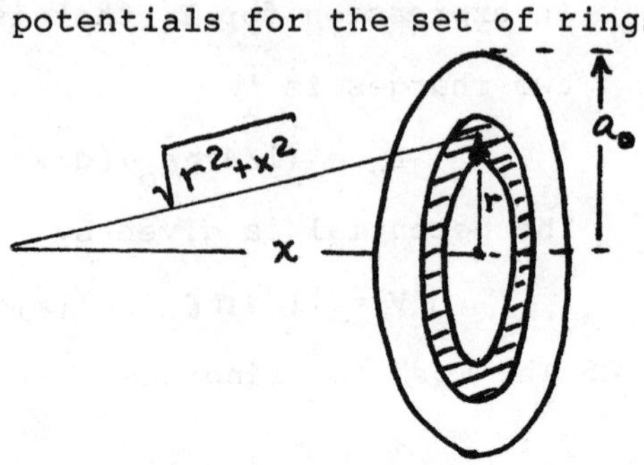

$$V = (\sigma/2\epsilon_o)\int_0^{a_o} rdr/[x^2 + r^2]^{\frac{1}{2}}$$

The indefinite integral is "easy".

$$\int rdr/[x^2 + r^2]^{\frac{1}{2}} = [r^2 + x^2]^{\frac{1}{2}},$$

BUT you must be careful in evaluating the result at the lower limit. Physically, the quantity $[x^2 + r^2]^{\frac{1}{2}}$ represents a distance and so is always positive. Remembering this we write

$$\int_0^{a_o} rdr/[x^2 + r^2]^{\frac{1}{2}} = [x^2 + a^2]^{\frac{1}{2}} \pm x$$

$+x$ for $x \leqslant 0$; $-x$ for $x \geqslant 0$.

The final result for V is

$$V = (\sigma/2\epsilon_o)\{[x^2 + a^2]^{\frac{1}{2}} \pm x\}; \quad +x \text{ for } x \leqslant 0,\ -x \text{ for } x \geqslant 0.$$

28.38 a) The electric field is related to V by $E = -dV/dx$. For the dipole,

$$V = (2aq/4\pi\epsilon_o)x^{-2}$$
$$E = -dV/dx = -(2aq/4\pi\epsilon_o)(-2x^{-3}) = (4aq/4\pi\epsilon_o)x^{-3}$$

b) Given the electric field we can integrate to determine the change in potential

$$V(x) - V(x_o) = -\int_{x_o}^{x} Edx$$

If we define $V(x_o) = 0$ at $x_o \rightarrow \infty$

$$V(x) = -\int_{\infty}^{x} Edx = -(4aq/4\pi\epsilon_o)\int_{\infty}^{x} dx/x^3 = (4aq/4\pi\epsilon_o)(\tfrac{1}{2}x^{-2})\Big|_{\infty}^{x}$$
$$= (2aq/4\pi\epsilon_o)x^{-2}$$

28.45 HINT: Let r_1 and r_2 denote the distances from the charges $+q$ and $-2q$ to an arbitrary point P. The potential at P is

$$V = (1/4\pi\epsilon_0)(q/r_1 - 2q/r_2)$$

In order that $V = 0$ we must have $r_2 = 2r_1$ for all points on the zero potential surface.

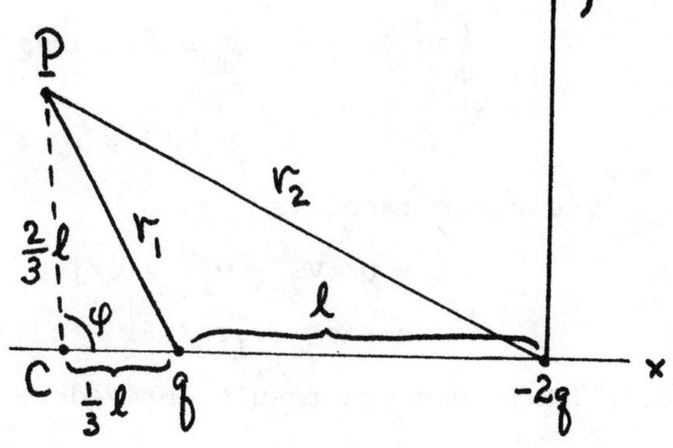

Use the law of cosines (See Appendix 4) to express r_1^2 and r_2^2 in terms of and the angle φ indicated in the figure. For points P located a distance $(2/3)\ell$ from the point C you will find $r_2^2/r_1^2 = 4$, showing that the zero potential surface is a sphere of radius $(2/3)\ell$ centered at $(-4\ell/3,0,0)$.

29.4 The charge on the capacitor is

$$Q = CV = (1\ F)(1000\ V) = 1000\ C$$

The density of Al is 2.7 gm/cm^3 so 1 cm^3 has a mass of 2.7 gm. The atomic weight of Al is 27, so 1 cm^3 consists of 0.1 mole or 6.02×10^{22} atoms. Each atom carries 13 electrons so there are $13(6.02 \times 10^{22})$ electrons in 1 cm^3 of Al. Each electron carries a negative charge of 1.60×10^{-19} C so the total electron charge in 1 cm^3 of Al is

$$Q_{Al} = 13(6.02 \times 10^{22})(1.60 \times 10^{-19}\ C) = 1.25 \times 10^5\ C$$

This is 125 times the charge on the capacitor.

29.10 The relation

$$-dV/dr = E = (Q/4\pi\epsilon_0)(1/r^2)$$

can be integrated to determine the potential difference.

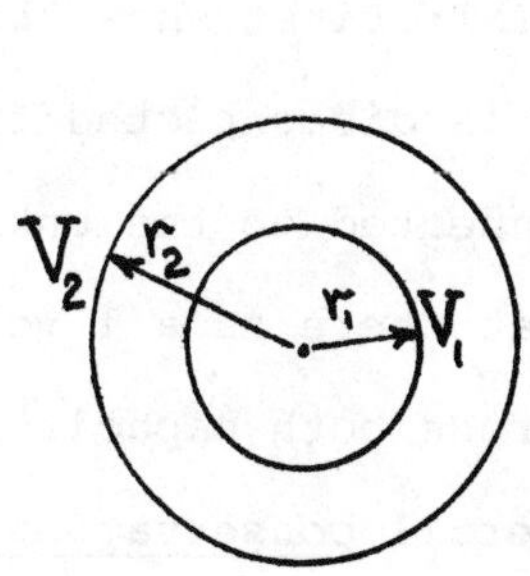

$$dV = -(Q/4\pi\epsilon_0)dr/r^2$$

Integrating from r_1 to r_2 gives the potential difference between the spheres

$$\int_{V_1}^{V_2} dV = V_2 - V_1 = -(Q/4\pi\epsilon_o)\int_{r_1}^{r_2} dr/r^2$$

$$= (Q/4\pi\epsilon_o)[1/r_1 - 1/r_2]$$

The capacitance is

$$C = Q/(V_2 - V_1) = Q/\{(Q/4\pi\epsilon_o)[1/r_1 - 1/r_2]\}$$

$$= 4\pi\epsilon_o\{r_1 r_2/(r_2 - r_1)\}$$

29.11 HINT: Use the result derived in Problem 29.10.

29.15 HINT: The variables x and y are related by

$$y = d + x\tan\theta$$

Note that

$$dx = dy/\tan\theta$$

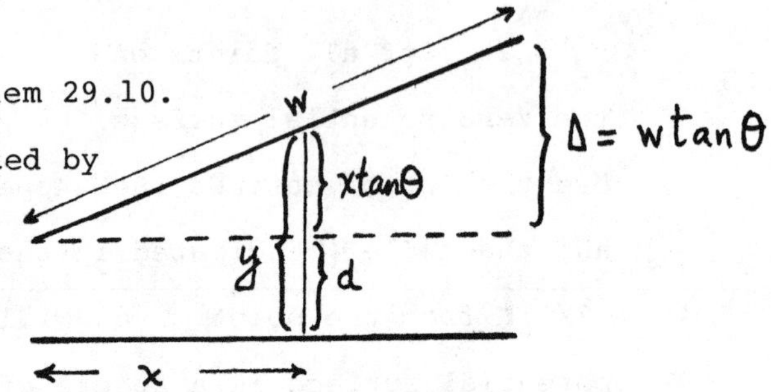

29.18 a) The parallel combination of 4 μF and 12 μF has an equivalent capacitance equal to their sum, 16 μF. Replacing the parallel combination by its equivalent 16 μF capacitance leaves a series combination of 16 μF and 6 μF. The equivalent capacitance for 6 μF in series with 16 μF is

$$C = (6)(16)/(6 + 16) = 4.36\ \mu F$$

b) The equivalent capacitance is given by

$$1/C = 1/4 + 1/6 + 1/12 = 6/12 = 1/2$$

so C = 2 μF.

c) The equivalent capacitance of capacitors in parallel is simply their sum:

$$C = 4 + 6 + 12 = 22\ \mu F$$

29.21 HINT: First show that the 5 μF capacitor carries a charge of 50 μC when it is disconnected from the 10-volt battery. When the 5 μF capacitor is then connected to the unknown capacitor the 50 μC is redistributed in such a way that there is a 1 volt potential difference across both capacitors - there being an overall <u>conservation of charge</u>.

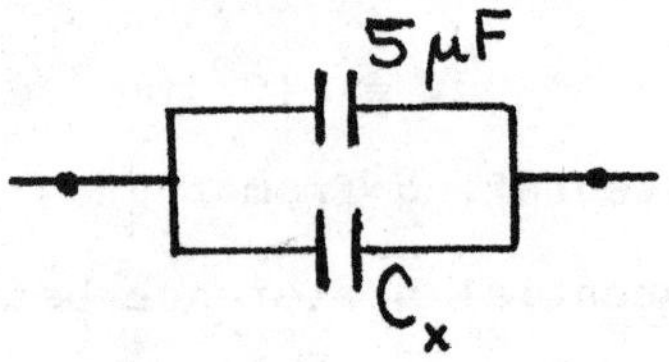

29.28 The capacitance of a parallel plate capacitor is given by

$$C = \epsilon_o A/d \; ; \quad A_1 = \text{plate area; } d = \text{plate separation}$$

For two capacitors in series, one with plate area A_1 and the other with plate area $A - A_1$, the equivalent capacitance is given by

$$1/C = d/\epsilon_o A_1 + d/\epsilon_o(A-A_1)$$

The maximum value of C corresponds to the minimum value of 1/C, which can be determined by requiring

$$d(1/C)/dA_1 = 0$$

Carrying out the differentiation,

$$d(1/C)/dA_1 = (d/\epsilon_o)\{ -1/A_1^2 + 1/(A - A_1)^2\} = 0$$

This gives

$$A_1^2 = (A - A_1)^2 \Rightarrow A_1 = A - A_1 \Rightarrow A_1 = \tfrac{1}{2}A$$

With $A_1 = \frac{1}{2}A$, both capacitors have equal plate areas of $\frac{1}{2}A$. This minimizes 1/C and maximizes C.

29.31 HINT: There are two useful relations: $U = Q^2/2C$; $U = \frac{1}{2}CV^2$. If you can determine the charge of the 4 μF capacitor then $U = Q^2/2C$ lets you evaluate U. If you can determine the potential difference across the 4 μF capacitor then $U = \frac{1}{2}CV^2$ lets you evaluate U.

29.33 HINT: $U = \frac{1}{2}CV^2 = Q^2/2C$

29.37 a) The capacitance is given by $C = \epsilon_o A/x$. Substituting this into $U = Q^2/2C$ gives the desired result

$$U = (Q^2/2\epsilon_o A)x$$

b) The force F is related to the energy U by $F = -dU/dx$, and to the pressure P by $F = PA$. Thus

$$F = PA = -dU/dx = -(Q^2/2\epsilon_o A)$$

$$P = -Q^2/2\epsilon_o A^2 = -\sigma^2/2\epsilon_o$$

where $\sigma = Q/A$ is the surface charge density. The electric field is $E = \sigma/2\epsilon_o$ so equivalent expressions for P are

$$P = -\tfrac{1}{2}\sigma E = -\tfrac{1}{2}\epsilon_o E^2$$

29.39 HINT: The <u>total</u> work performed when a charge Q is raised through a potential difference V is QV. One half of this work is stored in the capacitor as electric energy.

29.41 HINT: Consider a volume element in the form of a cylindrical shell of radius r, thickness dr, and length L. Its volume is $2\pi rdrL$ and the energy stored in it is $(\frac{1}{2}\epsilon_o E^2)(2\pi rdrL)$.

29.46 a) From $C = \epsilon_o A/d$ we have

$$A = Cd/\kappa\epsilon_o = (10^{-9})(5 \times 10^{-6})4\pi(8.99 \times 10^{9})/2.1$$

$$= 2.69 \times 10^{-4}\ m^2 = 2.69\ cm^2$$

b) The breakdown potential difference equals the product of the breakdown electric field and the plate separation

$$V_b = E_b d = (19 \times 10^3\ V/mm)(5 \times 10^{-3}\ mm) = 95\ V$$

29.54 From $C = \kappa\epsilon_o A/d$ we have the result

$$C_{teflon}/C_{air} = \kappa_{teflon}/\kappa_{air} = 2.1$$

because A and d are the same for both capacitors. From the relations

$$Q = CV; \quad U = \tfrac{1}{2}CV^2$$

we see that for equal potential differences (15 V here) the ratios Q_{teflon}/Q_{air} and U_{teflon}/U_{air} are both equal to the ratio of the capacitances:

$$Q_{teflon}/Q_{air} = 2.1; \quad U_{teflon}/U_{air} = 2.1$$

29.57 HINT: If the battery remains connected then the potential difference across the capacitor remains constant. If the battery is disconnected then the charge on the capacitor remains constant. The relations

$$U = \tfrac{1}{2}CV^2 \ ; \ U = Q^2/2C$$

are useful, as is $C = \kappa C_1$, where C_1 is the capacitance when the dielectric constant is unity.

29.62 HINT: The potential drop across the capacitor equals the sum of the drops across the slab and the air space - which indicates a series combination. Recall that the equivalent capacitance of two capacitors, C_1 and C_2, in series is

$$C = C_1C_2/(C_1 + C_2)$$

30.4 a) The current equals the charge Q that reaches the target in a time t divided by t.

$$I = Q/t$$

The charge Q equals the number of protons times the proton charge

$$Q = Ne \quad ; \; e = 1.60 \times 10^{-19}\ \text{C}$$

This gives $I = Ne/t$. The quantity N/t is the number of protons per second that reach the target

$$N/t = I/e = 10^{-3}\ \text{C/s}/(1.60 \times 10^{-19}\ \text{C}) = 6.25 \times 10^{15}/\text{s}$$

b) The number of protons in a 1 m length of the beam is given by

$$(\#\ \text{in 1 m}) = (\#/\text{s past any point})(\text{time to travel 1 m})$$

$$= (6.25 \times 10^{15}/\text{s})(1\ \text{m}/3 \times 10^{8}\ \text{m/s}) = 2.08 \times 10^{7}$$

30.8 NONRELATIVISTIC TREATMENT: a) The kinetic energy equals the electric work performed.

$$\tfrac{1}{2}mv^2 = qV$$

$$v = [2qV/m]^{\frac{1}{2}}$$

$$= [2(1.60 \times 10^{-19})(19 \times 10^{3})/(0.911 \times 10^{-30})]^{\frac{1}{2}}$$

$$= 8.17 \times 10^{7}\ \text{m/s}$$

b) $I/A = J = nqv$ gives

$$n = (I/A)/qv$$

$$= \{130 \times 10^{-6}/(\pi \times 10^{-6}/4)\}/(1.60 \times 10^{-19})(8.17 \times 10^{7})$$

$$= 1.27 \times 10^{13}/\text{m}^3$$

RELATIVISTIC TREATMENT: a) We can use the result derived in Problem 28.4.

$$v = c[2KE_o + K^2]^{\frac{1}{2}}/(E_o + K) \quad ; \; E_o = 511\ \text{keV},\ K = 19\ \text{keV}$$

$$v = 3 \times 10^{8}\ \text{m/s}[2(19)(511) + 361]^{\frac{1}{2}}/(511 + 19)$$

$$= 7.96 \times 10^{7}\ \text{m/s}$$

b) Following the same steps as in b) above, but with the speed $= 7.96 \times 10^{7}$ m/s gives $n = 1.30 \times 10^{13}$ m/s.

30.10 HINT: The current density $J = I/A$ is the same for wires with the same current-carrying capacity.

30.13 HINT: $Q = \int_0^\infty I\,dt$

30.15 The number of vehicles passing any point in time t equals the number contained in a length vt, where v is the vehicle speed. This number is nvt, where n is the number of vehicles per unit length. The rate at which vehicles pass is nvt/t = nv. The rate of mass flow is the mass per vehicle times the vehicle rate

rate of mass flow = Mnv

This result is analogous to $J = qnv_D$ for current density. Strictly speaking the expression Mnv is correct only if M and/or v is the same for all vehicles because the average of a product is not equal to the product of the averages.

A more mathematical treatment goes like this:

dm/dt = (dm/dx)(dx/dt) = (dm/dx)v

dm/dx = (mass/vehicle)(vehicles/length) = Mn

dm/dt = Mnv

30.16 Following the treatment above in Problem 30.15,

$I = dq/dt = (dq/dx)(dx/dt) = \lambda v$

30.22 HINT: $R = \rho \ell/A$ where $A = (\pi/4)(d_2^2 - d_1^2)$ = cross-sectional area of Cu.

30.29 HINT: The resistance is given by $R = \rho\ell/A$, where $\rho = 1.724 \times 10^{-8}\ \Omega\cdot m$ is the resistivity and $A = (\pi/4)(1.25 \times 10^{-3})^2 = 1.23 \times 10^{-6}\ m^2$ is the cross-sectional area of the wire. Evaluating the length of the wire is the challenging part of this problem. Write

ℓ = (# turns in each layer)(circumference of layer 1 + circumference of layer 2 + ... + ...layer 25)

The number of turns in each layer is 80 - you should explain why.

The circumference of successive layers increases. Let D = core diameter and let d = diameter of wire. Then

Circumference of layer 1 = $\pi(D + d)$

Circumference of layer 2 = $\pi(D + 3d)$

....

Circumference of layer 25 = $\pi(D + 49d)$

To evaluate the sum of circumferences recall $1+3+5+...+2N+1 = (N+1)^2$.

30.34 HINT: The resistance of a segment of length dx is given by

$$dR = \rho dx/A$$

Integrating dR from x = 0 to x = L gives the resistance of the full wire.

30.42 The length of the cycle is 5 s. For a uniform increase from 0 to 100 V in the "on" part of the cycle,

$$V = 100\ \text{V}\ (t/1\ \text{s}) \qquad 0 \leq t < 1\text{s}$$

$$V = 0 \qquad 1\text{s} < t \leq 5\text{s}$$

The average power, $\overline{P}$, equals the the total energy delivered in 5s divided by the cycle time of 5 s. The power is

$$P = V^2/R = (100)^2 t^2/20 \qquad 0 \leq t < 1\text{s}$$

$$P = 0 \qquad 1\ \text{s} < t \leq 5\ \text{s}$$

The average power is

$$\overline{P} = \int_0^{1\text{s}} P dt/(5\ \text{s}) = 500 \int_0^{1\text{s}} t^2 dt/(5\ \text{s}) = (500/15)\ \text{W} = 33.3\ \text{W}$$

30.55 HINT: Note that the cross-sectional area of the copper wire is that of a cylindrical shell with an inner diameter of 2 mm and an outer diameter of 4 mm. The current in each wire is simply the ratio of the potential drop across that wire to the resistance of that wire. The potential drop across each wire is 1 volt.

30.58 a) The maximum power is given by

$$P_{max} = VI_{max} = 12\ \text{V} \cdot 3\ \text{A} = 36\ \text{W}$$

b) With identical bulbs in parallel, each has the same resistance, each draws the same current, and the same power is delivered to each. At 5 W per bulb, a maximum of 7 bulbs is possible. This would require a power of 35 W and would draw (35/36) of the 3 A maximum current.

30.65 The power is given by $P = RI^2$. The current I is related to the current density J by $I = JA$. The resistance is $R = \rho \ell/A$. The power per unit volume is

$$P/A\ell = RI^2/A\ell = (\rho \ell/A)(JA)^2/A\ell = \rho J^2$$

30.66 HINT: Use the result derived in Problem 30.65.

30.67 HINT: Use $J = \sigma E$, and the result derived in Problem 30.65.

31.2 a) The chemical energy converted equals the electric work performed by the emf, which equals the emf ($\mathcal{E}$) times the charge delivered (q).

$$W = \mathcal{E}q$$

The charge delivered in a time t is q = It, so

$$W = \mathcal{E}It = (12\ V)(1\ A)(600\ s) = 7.2\ kJ$$

b) The energy produced in the resistor (E_R) is the product of the joule heating rate RI^2 and the time t.

$$E_R = RI^2t$$

As shown in part c) below, the resistance is 11 ohms, so

$$E_R = (11\Omega)(1\ A)^2(600\ s) = 6.6\ kJ$$

c) The current and emf are related by

$$\mathcal{E} = (r + R)I$$

Solving for R gives

$$R = (\mathcal{E}/I) - r = (12\ V/1\ A) - 1\Omega = 11\Omega$$

31.3 HINT: Let R_1 be the external resistance when the current is I and let R_2 be the external resistance when the current is 2I. Then

$$\mathcal{E}/(r + R_1) = I; \qquad \mathcal{E}/(r + R_2) = 2I$$

where r is the internal resistance and $\mathcal{E}$ = 12 V. This pair of equations contains four unknowns, r, I, R_1, and R_2, and so is incomplete. The information given concerning the power delivered to R_1 and R_2 supplies two more equations relating the unknowns - giving a total of four equations involving the four unknowns.

31.11 The equivalent resistance is given by

$$1/R = 1/R_1 + 1/R_2 + \ldots\ldots + 1/R_N > 1/R_{min}$$

where R_{min} denotes the smallest resistance in the combination. With $1/R > 1/R_{min}$ we have the desired result, $R < R_{min}$.

31.14 HINT: For the series connection

$$\mathcal{E} = (R_1 + R_2)I_{series}$$

For the parallel connection

$$\mathcal{E} = (R_1R_2/R_1+R_2)I_{parallel}$$

31.17 HINT: Bulb brightness is proportional to the power developed (RI^2).

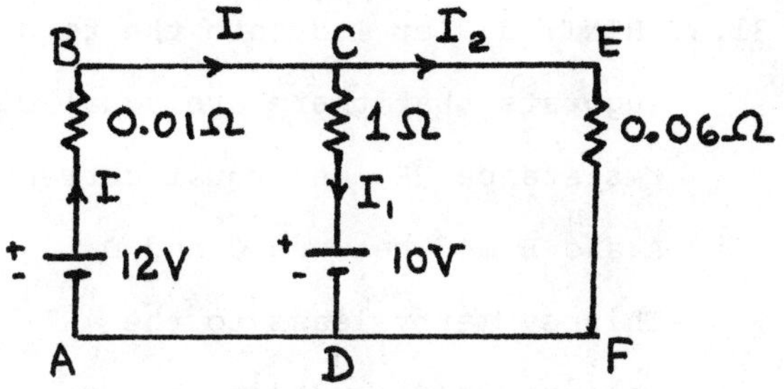

31.21 Let I denote the current delivered by the live battery and let I_1 and I_2 denote the currents leaving junction C in the figure. At C Kirchhoff's junction rule states

(1) $I = I_1 + I_2$

For the loop ABCDA Kirchhoff's loop rule states

(2) $12\ V - 0.01I - 1\ I_1 - 10\ V = 0$

For the loop CEFDC Kirchhoff's loop rule gives

(3) $-0.06I_2 + 10\ V + 1\ I_1 = 0$

Equations 1, 2, and 3 relate the three unknown currents. Using 1) in 2) eliminates I and gives

4) $2\ V = 1.01I_1 + 0.01I_2$

Solving 4) for I_1 and substituting into 3) gives

$$0.0699I_2 = 11.98\ V \Rightarrow I_2 = 171.4\ A$$

I_1 follows from 3).

$$I_1 = 0.06I_2 - 10 = 0.06(171.4) - 10 = 0.28\ A$$

31.29 HINT: For fixed values of $\mathcal{E}$ and r, the maximum power is determined by the condition dP/dr = 0. Note that there are three solutions of dP/dr = 0. Can you interpret all three?

31.36 HINT: a) R = V/I

b) The resistance R_{BC} between B and C is that of a 10 ohm resistor in parallel with a 30 ohm resistance. The resistance between A and D is the resistance of the complete network.

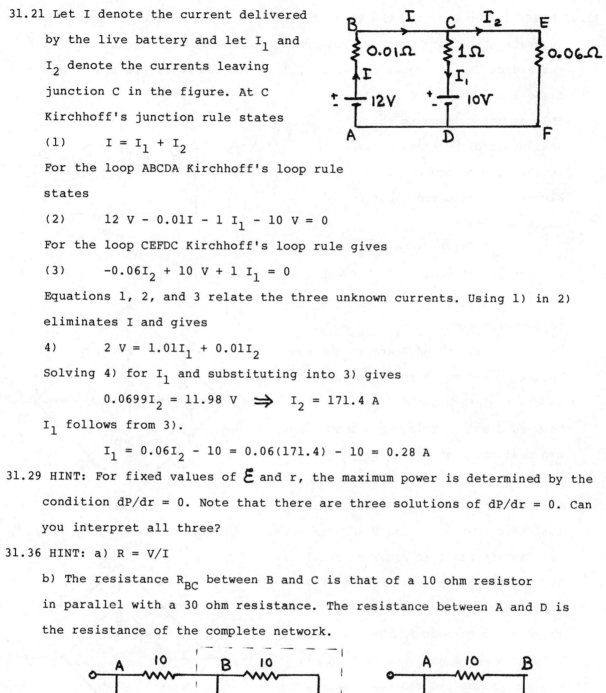

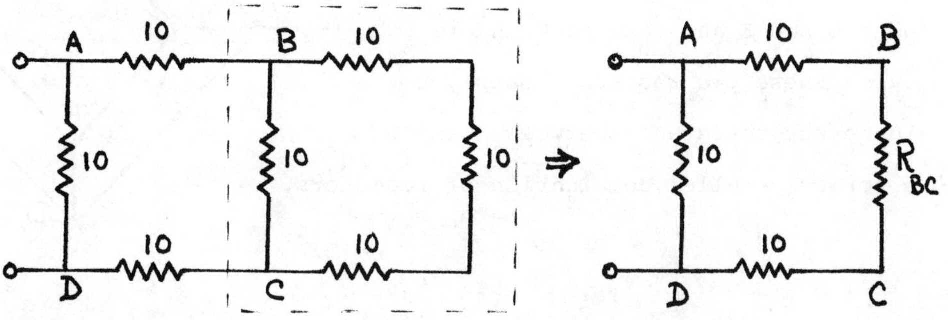

31.37 HINT: a) Let I denote the total current reaching junction A. Symmetry suggests that there are equal currents (I_1) in the two legs having a resistance 2R, and equal currents (I_2) in the two equal resistors between A and B and between C and D.

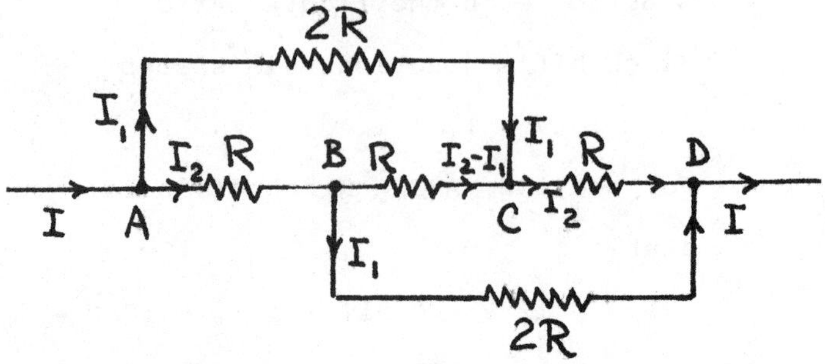

This symmetry leads to the assignment of currents shown in the figure. You can use Kirchhoff's loop and junction rules to find

$$I_1 = 2I/5; \; I_2 = 3I/5$$

The equivalent resistance between A and D is defined by

$$R_{equiv}I = V_{AD} = 2RI_1 + RI_2$$

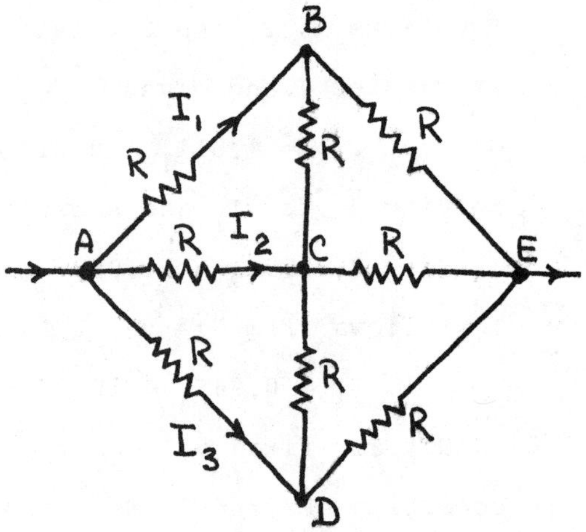

b) Symmetry requires $I_1 = I_2 = I_3$. This means B, C, and D are at the same potential and so there is no current in the resistors between B and C and between C and D. This means that these two resistors can be removed without altering the currents. Removing these two resistors lets you determine the resistance between A and E as a parallel combination of resistors.

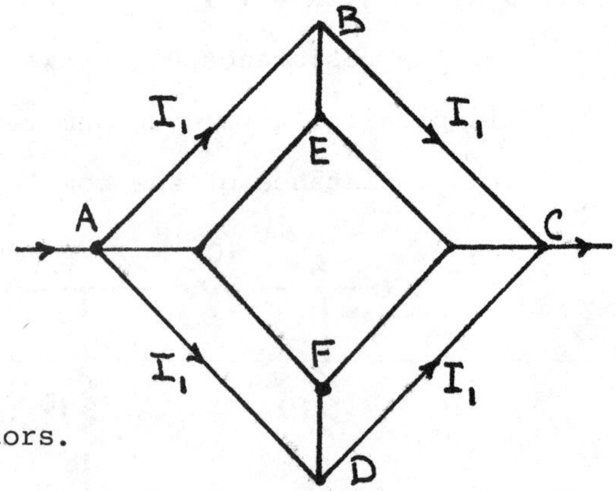

c) Symmetry requires equal currents (I_1) in the four outer legs. Kirchhoff's junction rule at B and D then shows that there is no current in the resistors between B and E and between D and F. Removing these two resistors lets you evaluate the resistance between A and C as a series-parallel combination of resistors.

31.40 HINT:

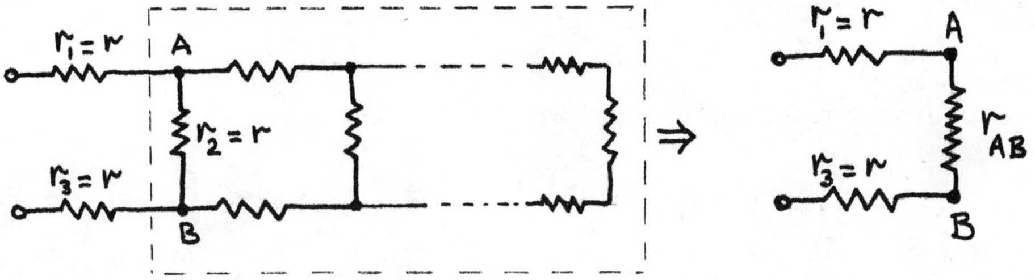

a) The network is equivalent to r_1 and r_3 in series with r_{AB} where r_{AB} is the equivalent resistance between A and B. The figure suggests that r_{AB} is equivalent to r_2 in parallel with the remaining portion of the infinite network - which has a finite resistance.

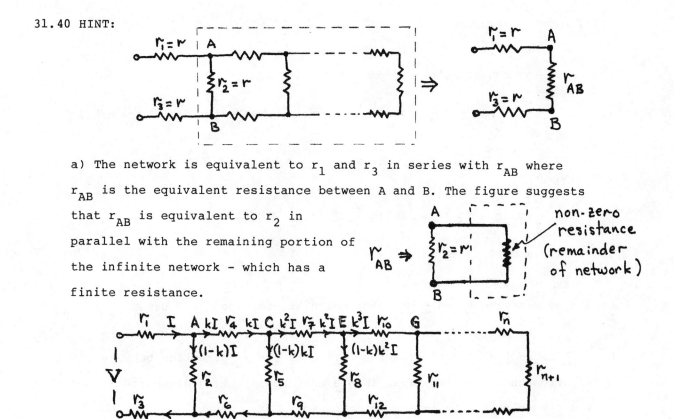

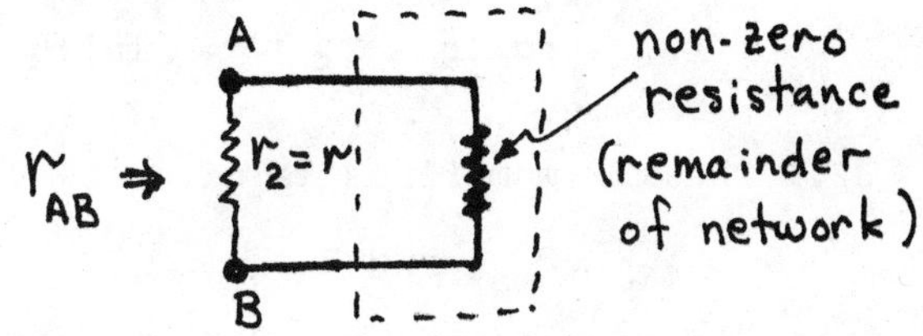

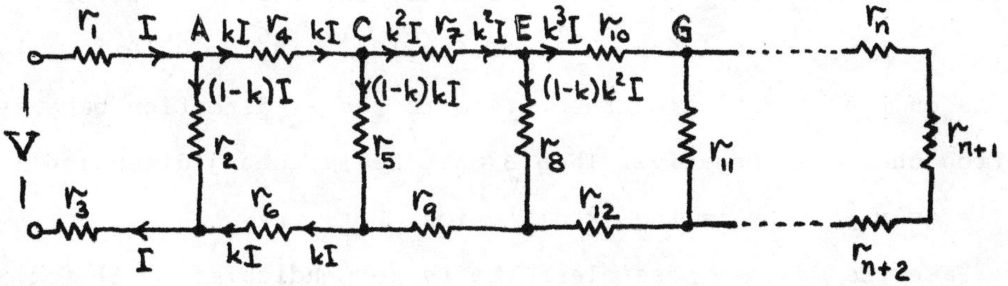

b) Exploit symmetry and envision what happens to the currents in all resistors if the potential difference across the left end is reversed.

c) Apply junction rule at A.

d) Because the network is <u>infinite</u> the junctions at C, E, G,.... are equivalent in that the <u>same fractions</u> of the current reaching these junctions leave along corresponding paths. Along the top of the network the fraction leaving each junction is k. A fraction 1-k leaves each junction along the vertical legs. This leads to the assignment of currents shown in the figure. See the Programs portion of this manual for a computer approach to this problem.

31.43 HINT: $U = q^2/2C$

31.46 HINT: $dq/dt = I$; $d\mathcal{E}/dt = 0$

31.48 The power developed by the emf is

$$P = \mathcal{E}I = (\mathcal{E}^2/R)e^{-t/RC}$$

The energy converted by the emf is the integral of P.

$$U = \int_0^\infty P dt = (\mathcal{E}^2/R)\int_0^\infty e^{-t/RC} dt = (\mathcal{E}^2/R)(RC)\underbrace{\int_0^\infty e^{-x} dx}_{=1} \quad (x = t/RC)$$

$$U = C\mathcal{E}^2$$

31.50 HINT:

$$\int_{q_0}^{q} dq/(q - C\mathcal{E}) = \ln[(q - C\mathcal{E})/(q_0 - C\mathcal{E})]$$

32.1 The maximum force is given by

$$F_{max} = qvB$$

and occurs when $\underset{\sim}{v}$ and $\underset{\sim}{B}$ are perpendicular. Solving for B gives

$$B = F_{max}/qv = 1.6 \times 10^{-14}/(1.6 \times 10^{-19}\ 10^6) = 0.10\ T$$

If $\underset{\sim}{v}$ is in the $+\underset{\sim}{i}$ direction then $q\underset{\sim}{v}$ is in the $-\underset{\sim}{i}$ direction because the electron charge is negative. Then $\underset{\sim}{B}$ must be in the $-\underset{\sim}{j}$ direction in order for $\underset{\sim}{F} = q\underset{\sim}{v} \times \underset{\sim}{B}$ to be in the $+\underset{\sim}{k}$ direction.

32.5 HINT: Take the maximum possible force ($\underset{\sim}{v}$ perpendicular to $\underset{\sim}{B}$) for your estimate.

32.6 HINT: a) The direction of $\underset{\sim}{B}$ must be such that the force is in the $+\underset{\sim}{j}$ direction. Remember that the electron charge is negative.

b) The average force $\overline{\underset{\sim}{F}}$ is defined by

$$\overline{\underset{\sim}{F}} = m\Delta\underset{\sim}{v}/\Delta t$$

where $\Delta\underset{\sim}{v}$ is the change in velocity in time Δt.

32.11 The maximum force is

$$F_{max} = qvB$$

Treating the motion nonrelativistically the kinetic energy is related to the accelerating potential V by

$$\tfrac{1}{2}mv^2 = qV \quad \Rightarrow \quad v = [2qV/m]^{\frac{1}{2}}$$

which gives

$$F_{max} = qB[2qV/m]^{\frac{1}{2}}$$

$$= (1.6 \times 10^{-19})(0.3)[2(1.6 \times 10^{-19})(10^4)/(0.911 \times 10^{-30})]^{\frac{1}{2}}$$

$$= 2.84 \times 10^{-12}\ N$$

32.13 HINT: With $v/c = 3/4$ you must use the relativistic equation for momentum. Equation 32.8 is replaced by

$$p = mv/(1 - v^2/c^2)^{\frac{1}{2}} = Bqr$$

If you evaluate r and find it is much smaller than one earth diameter you can conclude that the proton does not strike the earth (directly). If r is larger than one earth diameter you can conclude that the magnetic deflection is too slight to prevent the proton from hitting the earth.

32.16 HINT: Use Eq 32.7 and $\frac{1}{2}mv^2 = K$ to express B in terms of m, K, q, and r.

32.18 HINT: Use Eq 32.11. Correct to 3 significant figures the masses of ^{235}U and ^{238}U are simply 235u and 238u, where $u = 1.66 \times 10^{-27}$ kg.

32.26 HINT: The mass of the wire is given by

$$m = \rho A \ell$$

where ρ is its density, A is its cross-sectional area, and ℓ is its length.

32.32 In the diagram the voltage V measured between opposite faces of the sample includes a contribution IR due to the potential drop along the direction of the current. Let V_1 denote the measured potential drop when the magnetic field is in one direction, and let V_2 denote the measured potential drop when the magnetic field is reversed. Reversing the field changes the sign of the Hall voltage but does not affect the IR drop. Thus

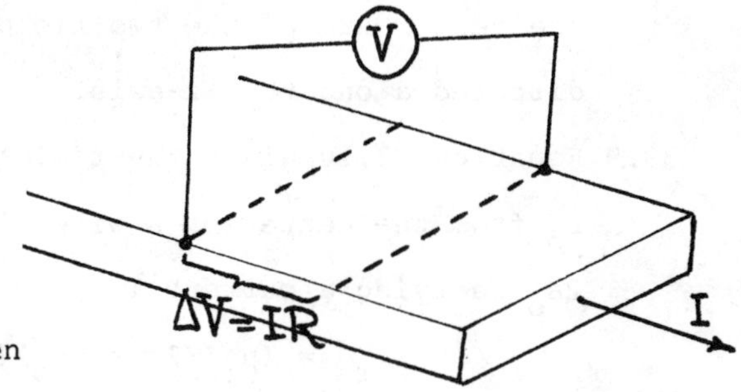

$$V_1 = V_H + IR; \quad V_2 = -V_H + IR$$

Subtracting equals gives

$$V_H = \frac{1}{2}(V_1 - V_2)$$

32.37 HINT: Use Eq 32.35, noting that $d\mathbf{B}/dx = 0.02x\mathbf{i}$ T/m

32.38 a) The torque is

$$\boldsymbol{\tau} = 0.0318(0.611\mathbf{j} - 0.792\mathbf{k}) \times (16.4\mathbf{j}) \text{ N·m} = 0.413\mathbf{i} \text{ N·m}$$

where we have used $\mathbf{j} \times \mathbf{j} = 0$, $-\mathbf{k} \times \mathbf{j} = \mathbf{i}$.

b) The potential energy is

$$U = -\boldsymbol{\mu} \cdot \mathbf{B} = -0.0318(0.611)(16.4) \text{ J} = -0.319 \text{ J}$$

where we have used $\mathbf{j} \cdot \mathbf{k} = 0$ and $\mathbf{j} \cdot \mathbf{j} = 1$.

33.3 Equation 33.10 gives the field a distance r_o from the center of a wire of length $2s_o$ carrying a current I.

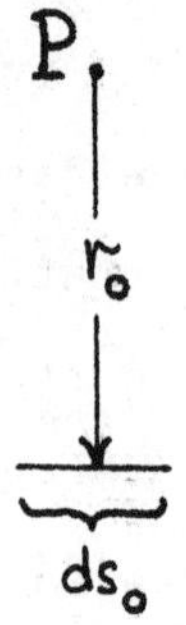

$$B = (\mu_o I/2\pi r_o)s_o/[s_o^2 + r_o^2]^{\frac{1}{2}}$$

In the limit $s_o \rightarrow \frac{1}{2}ds_o << r_o$ the factor

$$s_o/[s_o^2 + r_o^2]^{\frac{1}{2}} \rightarrow \frac{1}{2}ds_o/r_o$$

and the limiting form of Eq 33.10 is

$$B \rightarrow \mu_o I ds_o/2\pi r_o^2$$

For the geometry shown, $r \simeq r_o$ and $|\underset{\sim}{ds} \times \hat{\underset{\sim}{r}}| = ds_o$ and so Eq 33.1 becomes

$$dB = |\mu_o I\, \underset{\sim}{ds} \times d\hat{\underset{\sim}{r}}/4\pi r_o^2| = \mu_o I ds_o/4\pi r_o^2$$

in agreement with the limiting form of Eq 33.10.

33.6 HINT: The diagram shows the magnetic fields produced by two of the four sides. The vector sum of the two fields is directed along the +z-axis.

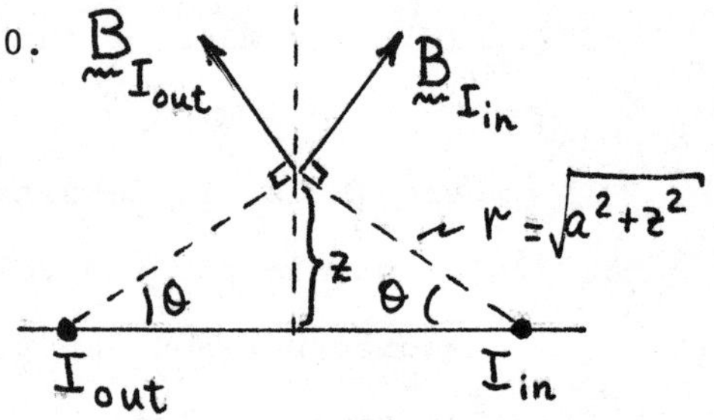

33.9 Equation 33.10 gives the field a distance r_o from the center of a wire of length $2s_o$ carrying a current I.

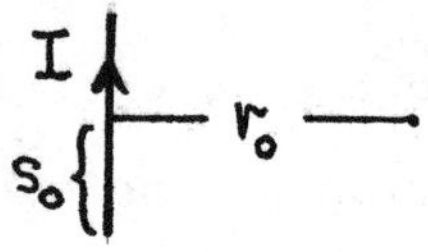

$$B_{wire} = (\mu_o I/2\pi r_o)s_o/[s_o^2+r_o^2]^{\frac{1}{2}}$$

The field at the center of a rectangular loop is the superposition of the fields of the four edges. Opposite edges produce equal fields at the center of the rectangle so

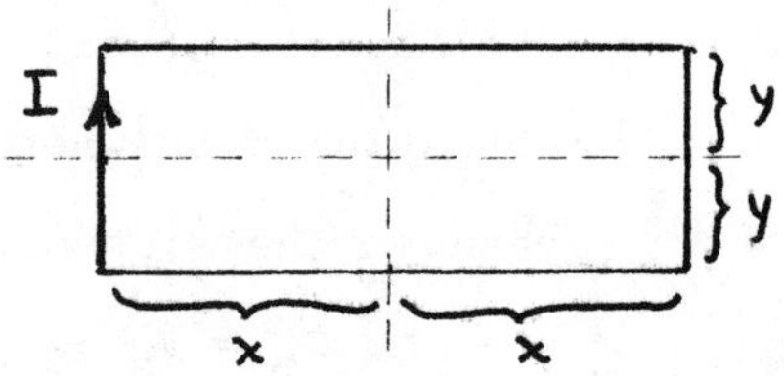

$$B = 2(\mu_o I/2\pi x)y/[x^2+y^2]^{\frac{1}{2}} + 2(\mu_o I/2\pi y)x/[x^2+y^2]^{\frac{1}{2}}$$
$$= (\mu_o I/\pi)[1/x^2 + 1/y^2]^{\frac{1}{2}}$$

Note that x and y enter this equation symmetrically. Interchanging x and y does not affect the value of B. The minimum value of B cannot single out either x or y as "special" because they are interchangeable. Symmetry therefore suggests that x = y for a minimum value of B. The choice x = y gives a square. To demonstrate analytically that x = y minimizes B we can write $4x + 4y = 4a$ = constant perimeter. Then $y = a-x$ and

$$B^2 = (\mu_o I/\pi)^2[1/x^2 + 1/(a-x)^2]$$

Setting $dB^2/dx = 0$ leads to the condition for minimum B^2 and for minimum B.

$$dB^2/dx = 0 = (\mu_o I/\pi)^2[-2/x^3 - 2/(a-x)^3]$$

which has as a solution $x = a-x = y$.

33.10 The magnetic field of the current strip can be evaluated by superposing the fields of infinitesimal-width strips. Let dI denote the current in the strip of width dy. For a uniform distribution of current

$$dI = Idy/2a$$

The field produced by dI at a distance R from the center of the strip is

$$dB = (\mu_o I/4\pi a)dy/R-y$$

Integrating from $y = -a$ to $+a$ gives the complete field

$$B = -(\mu_o I/4\pi a)\int_{-a}^{a} -dy/(R-y)$$

$$= -(\mu_o I/4\pi a)\ln[R-a/R+a]$$

$$= (\mu_o I/4\pi a)\ln[R+a/R-a]$$

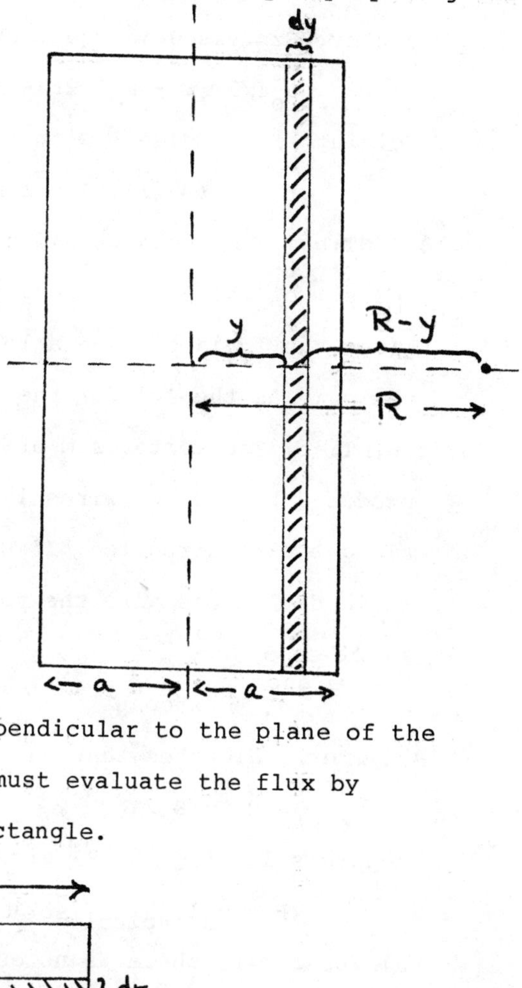

33.11 HINT: The field given by Eq 33.11 is perpendicular to the plane of the rectangle. Because B is not constant you must evaluate the flux by integrating B·Ldr over the area of the rectangle.

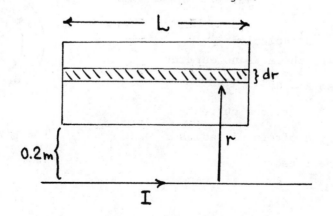

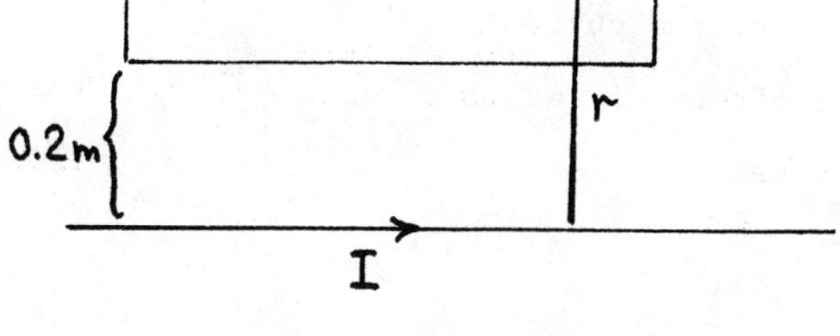

33.16 The field at the center of a circular loop of radius r is $\mu_o I/2r$ (Eq 33.15). The field a distance r from a long wire is $\mu_o I/2\pi r$ (Eq 33.11). The right-hand rule shows that the field of the wire and the outer loop add. The field of the inner loop opposes them and gives zero net field at the center. Expressed in equation form

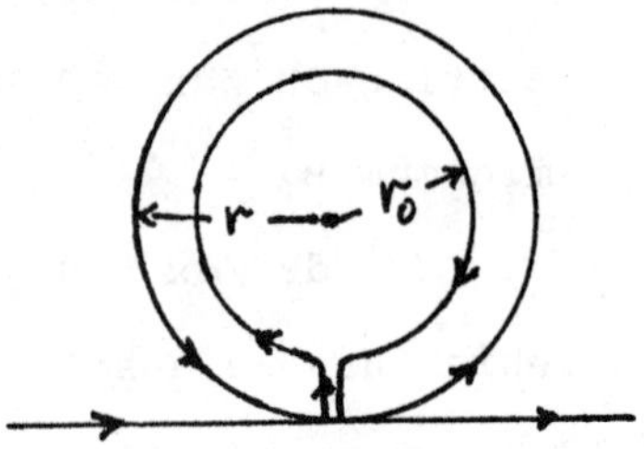

$$\mu_o I/2\pi r + \mu_o I/2r - \mu_o/2r_o = 0$$

Solving for r_o gives (with r = 1 m)

$$r_o = r(\pi/1+\pi) = 1\text{ m}(\pi/1+\pi) = 0.759\text{ m}$$

33.18 HINT: b) The field at a distance x from the midpoint between the coils is the sum of the fields produced by the two circular loops. Equation 33.14 gives the field of such a loop (but remember that the 'x' in Eq 33.14 is not the same as the 'x' for the Helmholtz coils).

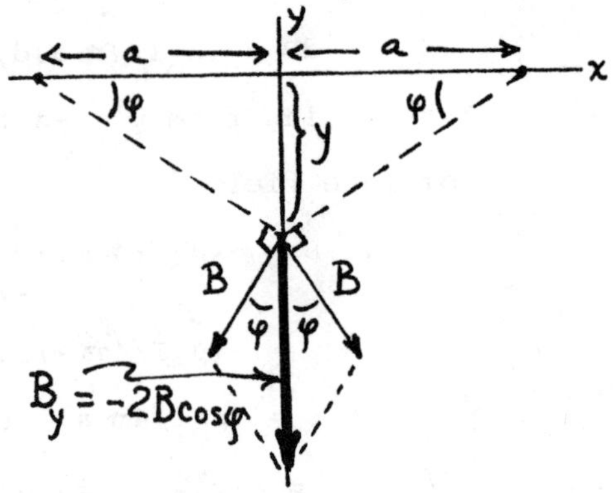

33.25 HINT: b) The vector sum of the fields produced by the two wires is directed along the $-\underset{\sim}{j}$ direction. If B denotes the field due to one wire the total field is given by

$$B_y = -2B\cos\varphi$$

33.29 Symmetry dictates that the field lines are circular and that

$$\oint \underset{\sim}{B}\cdot d\underset{\sim}{s} = B\cdot 2\pi r$$

Ampere's law (Eq 33.19) gives

$$B = \mu_o I_{encircled}/2\pi r$$

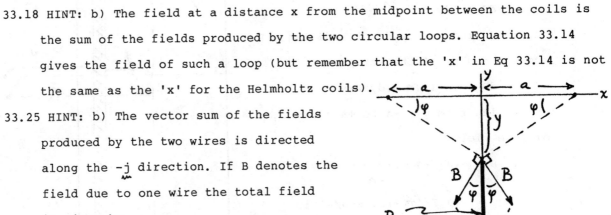

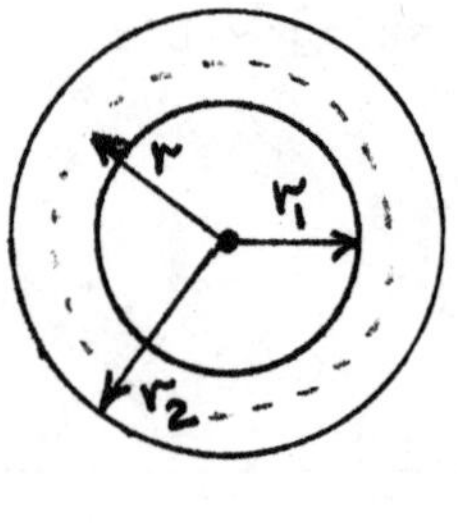

a) For $r < r_1$ there is no current encircled so B = 0.

b) For $r_1 < r < r_2$ the current encircled by a circle of radius r is given by the integral

$$J = I/\pi(r_2^2 - r_1^2)$$

$$I_{encircled} = \int_{r_1}^{r} J\ 2\pi r dr = [I/\pi(r_2^2 - r_1^2)]2\pi \int_{r_1}^{r} r dr$$

$$= I(r^2 - r_1^2)/(r_2^2 - r_1^2)$$

which gives

$$B = (\mu_o I/2\pi r)[(r^2 - r_1^2)/(r_2^2 - r_1^2)] \qquad r_1 < r < r_2$$

c) For $r > r_2$ the encircled current is I so

$$B = \mu_o I/2\pi r \qquad r > r_2$$

33.31 HINT: The field at the surface of a wire of radius a is given by

$$B = \mu_o I/2\pi a.$$

33.33

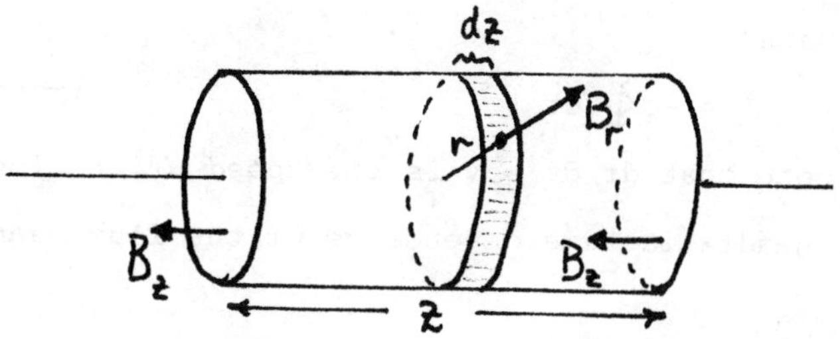

The flux for a closed surface is zero. The flux integral can be written as

$$\Phi_B = \int_{\text{Ends}} B_z \cdot dA + \int_{\text{Cylindrical Surface}} B_r\ 2\pi r dz = 0$$

There are no special points along the wire so B_z is a constant, independent of z - hence the fluxes through opposite ends cancel. There is no special direction around the wire. Hence any radial field B_r must be independent of direction and of z, so B_r is a constant over the surface of the cylinder.

$$\int B_r\ 2\pi r dz = B_r 2\pi rz$$

Thus,

$$\Phi_B = 0 + B_r 2\pi rz = 0 \Rightarrow B_r = 0$$

33.36 HINT: The magnetic field is given by Eq 33.27.

33.39 HINT: The length of wire on successive layers increases. See Problem 30.29 on page 72 of this manual.

33.46 HINT: The currents are parallel so the force is attractive. The "missing" portion is infinitesimal in length and so alters the field $\mu_o I/2\pi\Delta x$ acting on the displaced element by a negligible amount.

34.2 a) The emf is given by

$$\mathcal{E} = vB\ell = (4.31)(0.58)(0.82) = 2.05 \text{ V}$$

b) The potential difference equals the emf, so $\Delta V = \mathcal{E} = 2.05$ V

34.7 HINT: The field is spatially uniform so the flux threading the loop is given by $\Phi_B = B \cdot A$, where $A = \pi r^2$ is the loop area.

34.9 HINT: Refer to Problem 33.11, page 81 of this manual. By integrating $B\ell\,dx$ over the loop you can show that the flux threading the loop is

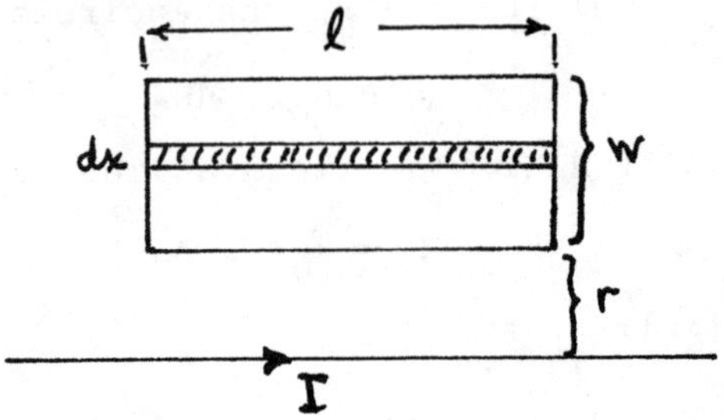

$$\Phi_B = (\mu_o I \ell / 2\pi)\ln(1 + w/r)$$

Evaluate

$$\mathcal{E} = -d\Phi/dt$$

and note that $dr/dt = v$ is the speed of the loop.

34.12 The quadratic time dependence of the flux means that Φ can be represented as

$$\Phi = a(t^2 - t_o^2); \quad t_o = 2 \text{ ms}$$

The constant a can be evaluated by making use of the fact that Φ equals 2×10^{-4} T•m^2 at t = 0. Setting $\Phi(t{=}0) = 2 \times 10^{-4}$ T•m^2 gives

$$a = \Phi(t{=}0)/t_o^2 = 2 \times 10^{-4} \text{ T•m}^2/(2 \times 10^{-3} \text{ s})^2$$
$$= 50 \text{ T•m}^2/\text{s}^2 = 50 \text{ V/s}$$

The flux is then

$$\Phi(t) = 50(t^2 - t_o^2) \text{ V/s, with t measured in seconds.}$$

The induced emf is given by

$$\mathcal{E} = -d\Phi/dt = -50\, d(t^2 - t_o^2)/dt = -100(t - t_o) \text{ V/s}$$

With $t_o = 2$ ms, $\mathcal{E} = 0.2$ V (t = 0); $\mathcal{E} = 0$ (t = 2 ms);

$\mathcal{E} = -0.2$ V (t = 4 ms)

34.18 We can evaluate E by making use of Faraday's law: $\mathcal{E} = -d\Phi/dt$.

The emf equals the line integral of $\underset{\sim}{E}$. The E-field lines are circular. If we evaluate the line integral around a circle of radius r

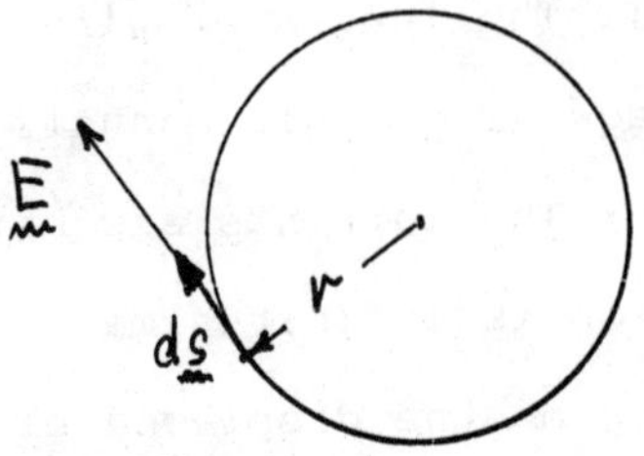

$$\mathcal{E} = \oint \underset{\sim}{E} \cdot d\underset{\sim}{s} = E \cdot 2\pi r$$

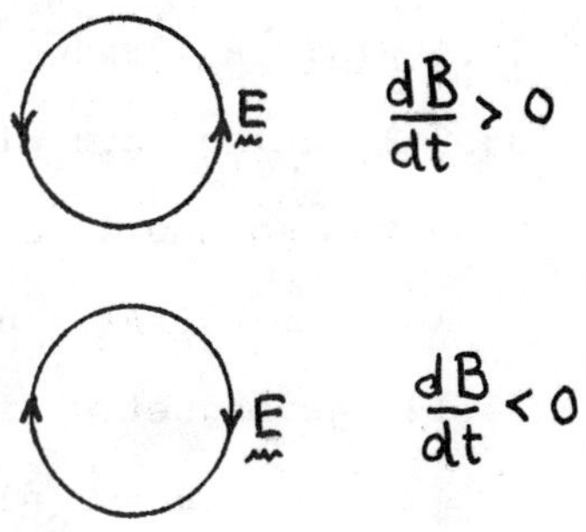

For $r \geqslant R$ the flux threading the loop defining the path of the line interal is $\Phi = B\pi R^2$

and

$$-d\Phi/dt = -\pi R^2 dB/dt$$

Faraday's law then gives (we ignore the minus sign)

$$E = (\pi R^2/2\pi r)dB/dt = (R^2/2r)dB/dt$$

The ignored minus sign is related to the direction of the induced electric field. For example, if an increasing magnetic field ($dB/dt > 0$) results in the electric field lines having a counterclockwise sense, then a decreasing magnetic field ($dB/dt < 0$) results in electric field field lines with a clockwise sense.

If $r \leqslant R$ the flux threading the loop is $\Phi = B\pi r^2$ and

$$-d\Phi/dt = -\pi r^2 dB/dt$$

and the induced electric field is

$$E = (\pi r^2/2\pi r)dB/dt = \tfrac{1}{2}r dB/dt$$

34.25 HINT: Use Eqs 34.16 and 34.15.

34.27 The power developed is related to the resistance and current by

$$RI^2 = P \Rightarrow I = [P/R]^{\frac{1}{2}} = [50/200]^{\frac{1}{2}} = 0.50 \text{ A}$$

34.28 a) The flux threading the loop can be expressed as

$$\Phi = BA \cdot \sin\omega t$$

where ωt is the angle between $\underset{\sim}{B}$ and the normal to the loop.

The maximum flux is

$$\Phi_{max} = BA = (0.682 \text{ T})(0.12 \text{ m})(0.37 \text{ m}) = 0.0303 \text{ T}\cdot\text{m}^2$$

b) The time rate of change of flux is

$$d\Phi/dt = \omega BA \cos\omega t$$

which has as its maximum value

$$(d\Phi/dt)_{max} = \omega BA = 2\pi\cdot 60(0.0303) \text{ T}\cdot\text{m}^2/\text{s} = 11.4 \text{ T}\cdot\text{m}^2/\text{s} = 11.4 \text{ V}$$

34.31 HINT: The flux threading the coil can be expressed as

$$\Phi = NBA \cdot \cos\omega t$$

where N is the number of turns, A is the coil area, and ω is the radian frequency of rotation.

35.5 HINT: Rearrange $\mathcal{E} = -LdI/dt$ as $dI = -(\mathcal{E}/L)dt$ and integrate to get I.

35.9 Equation 35.8 gives the solenoid inductance, $L = \mu_o N^2 A/\ell$. The volume of the solenoid is $A\ell$, so the inductance per unit volume is

$$L/A\ell = \mu_o N^2 \not{A}/\not{A}\ell^2 = \mu_o n^2 \quad ; \; n = N/\ell$$

35.12 The magnetic field of a solenoid is given by Eq 33.27

$$B = \mu_o nI$$

The number of turns per meter is

$$\mu_o = (B/I)/\mu_o = 0.1338/4\pi \times 10^{-7} = 1.07 \times 10^5 \text{ turns/m}$$

35.17 Both V and $\mathcal{E}$ have the same dimensions, so their ratio is dimensionless. Forming $[V]/[\mathcal{E}]$ gives

$$[V]/[\mathcal{E}] = [\not{I}R]/[Ld\not{I}/dt] = [t]/[L/R]$$

which shows that L/R has the dimensions of time.

35.18 HINT: Use Eq 35.13.

35.20 a) Kirchhoff's loop equation is

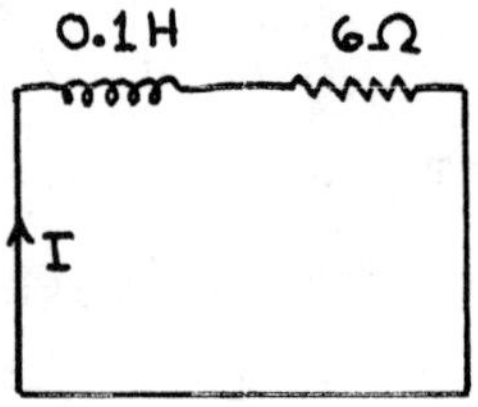

$$LdI/dt + RI = 0$$

The potential drop across the inductor is

$$-LdI/dt = RI = (6\,\Omega)(5\text{ A}) = 30\text{ V}$$

b) Solving for dI/dt gives

$$dI/dt = -RI/L = -30/0.1 = -300\text{ A/s}$$

c) We can estimate the current as the circuit is being broken by setting $dI/dt = \Delta I/\Delta t = (0-5)A/10^{-3}\text{ s} = -5000\text{ A/s}$.

$$I \simeq -(L/R)\Delta I/\Delta t = -(0.1/6)(-5000) = 83\text{ A}$$

How can the current be 83 A when it decreases from 5 A to zero? I'm glad you asked that question! Remember Lenz's law? As the circuit is being broken the induced emf reacts to prevent a flux change. The induced emf produces a brief surge in the current - it increases sharply before dropping to zero.

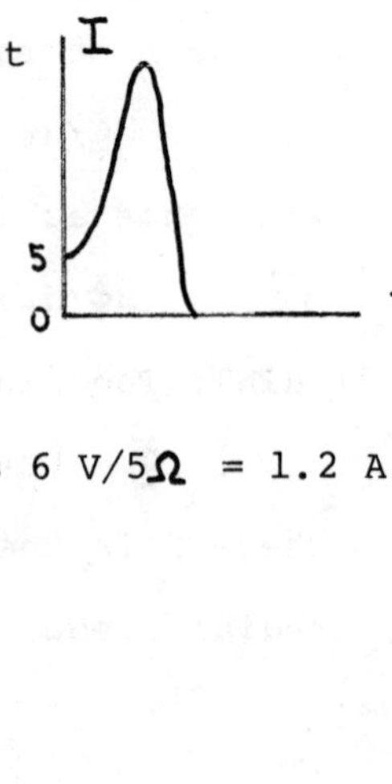

35.21 HINT: When the circuit is first opened, the current is $6\text{ V}/5\Omega = 1.2\text{ A}$.

35.24 HINT: Remember Lenz's law.

35.30 a) The magnetic energy density is given by Eq 35.23

$$u_M = B^2/2\mu_o = (4.5)^2/8\pi \times 10^{-7} = 8.06 \times 10^6 \text{ J/m}^3$$

b) The magnetic energy stored in the field equals u_M times the volume of the solenoid (the volume in which B is non-zero).

$$U_M = u_M A\ell = (8.06 \text{ J/cm}^3)\{26 \text{ cm}\cdot\pi(3.1 \text{ cm})^2\} = 6.32 \text{ kJ}$$

35.33 The inductance per unit length is given by Eq 35.11

$$L/\ell = (\mu_o/2\pi)\ln(r_2/r_1)$$

The magnetic energy is $U_M = \frac{1}{2}LI^2$, so the magnetic energy per unit length is

$$U_M/\ell = \tfrac{1}{2}(L/\ell)I^2 = (\mu_o/4\pi)\ln(r_2/r_1)I^2$$

35.38 HINT: The magnetic energy stored in a cylindrical shell of volume $2\pi rdr\ell$ is $(B^2/2\mu_o)2\pi rdr\ell$.

36.1 a) The empty solenoid field is given by Eq 36.4,

$$B_E = \mu_o(N/\ell)I = 4\pi \times 10^{-7}(4000)1 = 5.03 \times 10^{-3} \text{ T}$$

b) The change in the field when the solenoid is filled with copper is

$$\Delta B = \chi_m B_E = -10.8 \times 10^{-6}(5.03 \times 10^{-3} \text{ T}) = -5.43 \times 10^{-8} \text{ T}$$

The minus sign indicates that the field decreases when the diamagnetic copper is inserted.

36.3 HINT: $B = (1 + \chi_m)B_E$ gives $\chi_m = (B/B_E) - 1$. The empty solenoid field strength is given by Eq 36.4.

36.4 HINT: The induced field is

$$B_{induced} = \chi_m B_E$$

The induced field is related to the induced dipole moment per unit volume ($\mu_{induced}/V$) by

$$B_{induced} = \mu_o(\mu_{induced}/V)$$

36.8 HINT: Use Eq 36.11, $B_E = \mu_o\mu/V$.

36.13 If the relation $\chi_m = a\mu$ is substituted into Eq 36.14 the relation between B and μ is

$$B = (\mu_o/V)\{\mu + a\mu^2\}$$

A more fundamental approach produces a different result. The magnetic moment of an empty solenoid is proportional to the current in the solenoid coils, and so the induced moment (μ_I) is also proportional to the solenoid current. Eq 36.13 ($\mu_I = \chi_m \mu$) recognizes these proportionalities by introducing the magnetic susceptibility. Eq 36.13 is adequate when χ_m is constant. A more fundamental form of the relation is required when χ_m is not constant, namely

$$d\mu_I = \chi_m d\mu$$

This form recognizes the step-by-step nature of induction. If χ_m = constant, integration gives $\mu_I = \chi_m \mu$. But, with $\chi_m = a\mu$,

$$\int d\mu_I = \mu_I = \int a\mu d\mu = \tfrac{1}{2}a\mu^2 = \tfrac{1}{2}\chi_m \mu$$

and the relation between B and μ is

$$B = (\mu_o/V)\{\mu + \tfrac{1}{2}a\mu^2\}$$

36.18 HINT: The magnetic field at the center of a ring of radius r is given by

$$B = \mu_o I/2r$$

36.22 HINT: Make use of Eqs 36.11, 36.13, and 36.17.

36.27 From $B = (1 + \chi_m)B_E$ we have

$$\chi_m = (B/B_E) - 1$$

With $B = 7.0 \times 10^{-4}(B_E/\mu_o)$

$$\chi_m = (7.0 \times 10^{-4}/4\pi \times 10^{-7}) - 1 \simeq 556$$

With a two-significant figure input datum we round this to $\chi_m = 560$.

37.1 The angle between $\underset{\sim}{B}$ and the normal to the loop can be expressed as $\theta = \omega t$.

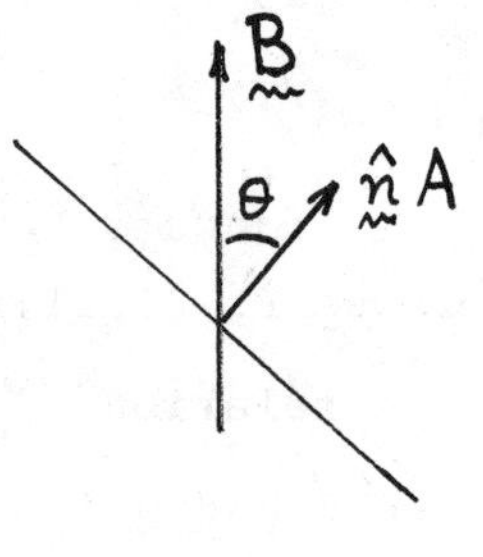

The flux threading the loop is

$$\Phi = \underset{\sim}{B}\cdot\hat{\underset{\sim}{n}}A = BA\cdot\cos\omega t$$

The induced emf follows from Faraday's law

$$\mathcal{E} = -d\Phi/dt = \omega BA\cdot\sin\omega t = \omega BA\cdot\sin\theta$$

37.5 The maximum current is

$$I_m = V_{max}/R = \sqrt{2}\cdot 115\ V/23\Omega = 5\sqrt{2}\ A$$

The current can be expressed as

$$I = I_m \cos\omega t \quad ; \quad \omega = 2\pi\nu = 377\ rad/s$$

The power is

$$P = RI^2 = RI_m^2\cos^2 377t$$

The minimum power is zero. The maximum power is

$$P_{max} = RI_m^2 = 23(5\sqrt{2})^2 = 1.15 \text{ kW}$$

The average power is given by Eq 37.6,

$$P_{av} = \tfrac{1}{2}RI_m^2 = 575 \text{ W}$$

37.7 HINT: The rms current is related to the maximum current by

$$I_{rms} = I_{max}/\sqrt{2}$$

37.10 HINT: The average potential is defined by

$$V_{av} = \int_0^{2\pi} V\, d(\omega t) \Big/ \int_0^{2\pi} d(\omega t)$$

Note also that because $V = \mathcal{E}|\sin\omega t|$,

$$\int_0^{2\pi} V\, d(\omega t) = 2\int_0^{\pi} V\, d(\omega t)$$

37.11 HINT: The current is related to the maximum emf ($\mathcal{E}$) by Eq 37.9,

$$I = -\omega C\mathcal{E}\sin\omega t$$

37.17 The current is given by Eq 37.9,

$$I = -\omega C\mathcal{E}\sin\omega t$$

The maximum current is proportional to $\omega = 2\pi\nu$. Increasing the frequency by a factor of 10^5 also increases the maximum current by the same factor.

37.22 HINT: $X_L = \omega L = 2\pi\nu L$

37.28 The current satisfies $L\,dI/dt = \mathcal{E}\sin\omega t$. Noting that

$$d(-\cos\omega t/\omega) = \sin\omega t$$

shows

$$dI/dt = d(-\mathcal{E}\cos\omega t/\omega L)$$

and hence that the steady state current is

$$I = (-\mathcal{E}/\omega L)\cos\omega t$$

37.31 HINT: The period is 6 ms so

$$\omega = 2\pi/T = 2\pi/0.006 \text{ s} = 1000\pi/3 \text{ rad/s}$$

The current is related to ω and L by Eq 37.15.

37.36 HINT: The small angle ($\varphi << 1$) approximation $\tan\varphi \simeq \varphi$ has as its arc tan counterpart

$$\tan^{-1}\varphi \simeq \varphi$$

37.39 HINT: The rms potentials across the resistor and capacitor are given by Eqs 37.42 and 37.43. Note that $\mathcal{E}/\sqrt{2}$ is the rms potential of the generator (10 V in this problem).

37.44 HINT: Use $Z^2 = R^2 + (\omega L)^2$; $\omega = 2\pi\nu$

37.46 The rms current is given by

$$I_{rms} = V_{rms}/Z$$

The impedance follows from

$$Z^2 = R^2 + [1/\omega C - \omega L]^2$$

$$= 10^2 + [(1/2\pi\cdot 100\cdot 10^{-5}) - 2\pi\cdot 100\cdot 1.5 \times 10^{-3}]^2$$

This gives

$$Z = 159\,\Omega$$

With $V_{rms} = 25$ V

$$I_{rms} = 25\text{ V}/159\,\Omega = 0.158\text{ A}$$

37.53 The thermal power is $P = RI_m = R\,\mathcal{E}^2_{rms}/Z^2$. At resonance Z = R and

$$P_{res} = R\,\mathcal{E}^2_{rms}/R^2 = \mathcal{E}^2_{rms}/R = (100)^2/200 = 50\text{ W}$$

The resonant radian frequency is

$$\omega_r = [1/LC]^{\frac{1}{2}} = [1/10^{-4}\cdot 10^{-8}]^{\frac{1}{2}} = 10^6\text{ rad/s}$$

At a frequency of $\omega = 0.9\omega_r = 9 \times 10^5$ rad/s,

$$Z^2 = R^2 + [1/\omega C - \omega L]^2$$

$$= (200)^2 + [(1/9\text{x}10^5\cdot 10^{-8}) - (9\text{x}10^5\cdot 10^{-4})]^2$$

$$= 40{,}400\,\Omega^2$$

The power at this frequency is

$$P = 200\cdot(100)^2/40{,}400 = 49.5\text{ W}$$

37.56 HINT: Remember that the half-maximum frequency width is the difference of the frequencies at which the power is half of its maximum value. Eq 37.75 gives the radian frequency difference.

37.64 HINT: Devices made to operate at 60 Hz will also operate at 50 Hz so the transformer need not change the frequency. It need only reduce the voltage from 240 V to 120 V. What ratio of primary:secondary turns is needed?

38.5 HINT: Use $\nu = \omega/2\pi$ and $\lambda = 2\pi/k$. One of the standard forms of a plane sinusoidal wave is given by Eq 38.8.

38.9 The dimensions of ordinary current are

$$[I] = \text{charge/time}$$

The dimensions of Maxwell's displacement current are

$$[I_{displ}] = [\epsilon_o d(\int \underset{\sim}{E} \cdot d\underset{\sim}{A})/dt] = [\epsilon_o(\text{electric flux/time})]$$

From Gauss' law we can see that electric flux has the dimensions of (charge/ϵ_o): $\Phi_E = \int \underset{\sim}{E} \cdot d\underset{\sim}{A} = q/\epsilon_o$. Thus

$$[I_{displ}] = [\epsilon_o(\text{charge}/\epsilon_o)/\text{time}] = \text{charge/time} = [I]$$

38.11 Equations 38.20 and 38.21 give the magnitudes and units of ϵ_o and μ_o.

$$\epsilon_o = 8.85419 \times 10^{-12}\ C^2/N\cdot m^2 \qquad (38.20)$$

$$\mu_o = 1.25664 \times 10^{-6}\ N/A^2 \qquad (38.21)$$

The units of $1/[\epsilon_o\mu_o]^{\frac{1}{2}}$ are (we set 1 C = 1 A•s)

$$1/[(C^2/\cancel{N}\cdot m^2)(\cancel{N}/A^2)]^{\frac{1}{2}} = 1/[\cancel{A^2}s^2/m^2)/\cancel{A^2})]^{\frac{1}{2}} = m/s$$

38.17 HINT: a) The intensity equals the power per unit area. b) Eq 38.35 relates intensity and energy density.

38.18 HINT: Equation 38.44 relates intensity and E-field amplitude.

38.20 The power is given by P = I A. If 60% of the incident power is absorbed,

$$P_{abs} = (0.60)(1.38 \times 10^3\ W/m^2)(10^6\ m^2) = 8.28 \times 10^8\ W$$

Not all of the absorbed energy could be converted into electric energy. In any sort of heat engine that converts heat into electric energy some of the absorbed energy would have to be ejected in accordance with the second law of thermodynamics.

38.25 The integral in the denominator is easy: $\int_0^{2\pi} d\theta = 2\pi$. You can work out or look up $\int_0^{2\pi} \cos^2\theta\, d\theta$. But, here's an opportunity to use the <u>numerical integration</u> technique, as developed in the Computer Programs section of this manual.

38.28 HINT: Each of the three polarizers transmits a plane-polarized wave. The plane of polarization is rotated by 30° by each polarizer.

39.5 From the figure,

$$d = r \sin(\theta_1 - \theta_2)$$

and

$$r = t/\cos\theta_2$$

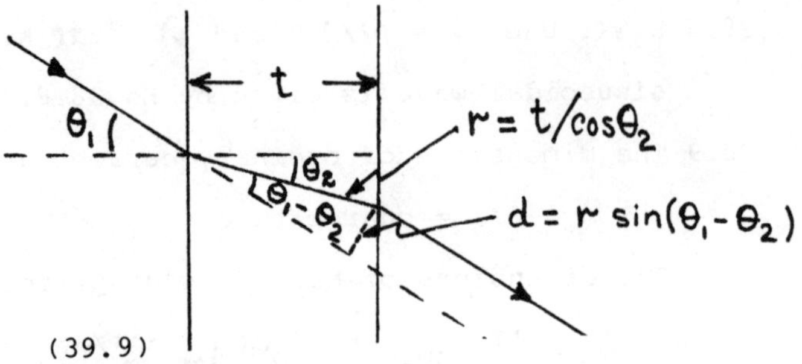

so

$$d = t\cdot\sin(\theta_1-\theta_2)/\cos\theta_2 \qquad (39.9)$$

39.10 Using Snell's law at the first interface with $i = 20^\circ$ and $n = 1.55$ gives

$$\sin(r) = \sin 20^\circ/1.55 = 0.2207$$

$$r = 12.75^\circ$$

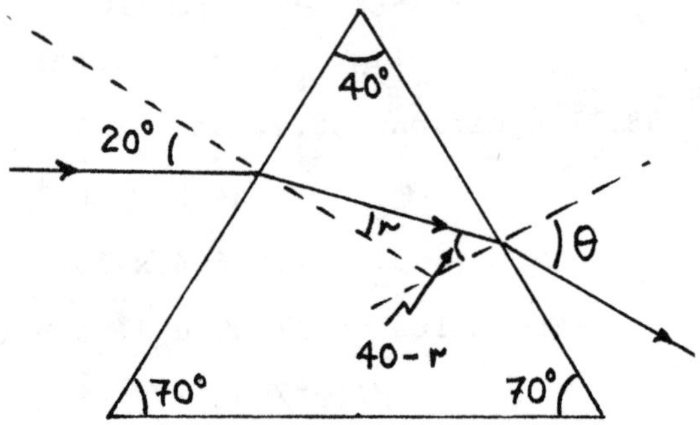

The angle of incidence at the second interface is $40-r = 27.25^\circ$. Using Snell's law at the second interface gives

$$\sin\theta = 1.55\cdot\sin 27.25^\circ = 0.7097$$

from which we find $\theta = 45.2^\circ$.

39.11 HINT: Label the angles of incidence and refraction and use the diagram to relate h and h_A to the tangents of those angles. Then use the fact that the angles are small to make the approximation, $\tan\theta \simeq \sin\theta$. Then use Snell's law to eliminate the sines.

39.20 Solving the mirror equation $1/s + 1/s' = 1/f$ for the image distance s' gives

$$s' = sf/(s-f)$$

The focal length is $f = \frac{1}{2}r = 0.6$ m. With an object distance $s = 5$ m,

$$s' = 5(.6)/(5-0.6) = 0.682 \text{ m}$$

You could not see your image in the mirror. The image is real - it could be displayed on a screen placed 0.682 m in front of the mirror. The type of image one sees in a mirror is a virtual image. These virtual images appear to be formed behind the mirror surface. Mathematically, such virtual images correspond to negative values of s'.

39.23 HINT: Remember $f = \frac{1}{2}r$ for a spherical mirror.

39.28 HINT: The image formed by the first lens becomes the object for the second lens.

39.33 HINT: You must use Eq 39.34 to determine f.

39.39 HINT: The cornea-to-retina distance of 2.12 cm is the focal length of the eye for an object at infinity.

40.3 Using Eq 40.8,

$$\lambda = a\sin\theta/m = (0.26\ \text{mm})(\sin 0.14^\circ)/1 = 6.35 \times 10^{-4}\ \text{mm} = 635\ \text{nm}$$

40.7 HINT: Use Eq 40.8 to determine the angle θ.

40.8 HINT: The two criteria for coherence are given on page 785.

40.13 Equation 40.57 defines the condition for constructive interference of the reflected light.

$$2nd = (m + \tfrac{1}{2})\lambda \qquad m = 0,1,2,...$$

m = 0 gives the minimum value of d.

$$d_{min} = \lambda/4n = 500\ \text{nm}/4(1.41) = 88.6\ \text{nm}$$

40.18 HINT: The phase difference between waves reflected at the top and bottom of the air wedge is

$$\pi + 2\pi(2d/\lambda)$$

For constructive interference this phase difference must be an integral multiple of 2π.

40.22 HINT: Remember that when the mirror moves a distance d, the optical path length changes by 2d.

41.2 From Eq 41.17 the intensity is zero when $\Phi = \pi$. From Eq 41.11

$$\Phi = (\pi D\sin\theta/\lambda) = \pi \Rightarrow \qquad \sin\theta = \lambda/D = 1/10$$

For $\sin\theta = 0.10$, the small-angle approximation $\sin\theta \simeq \theta$ is adequate, so

$$\theta = 0.1\ \text{radian}$$

41.4 From Eq 41.17,

$$I_\theta/I_o = (\sin\Phi/\Phi)^2; \qquad \Phi = \pi D\sin\theta/\lambda$$

The maxima other than the $\theta = 0$ central maximum occur <u>approximately</u> at

$$\Phi = \pm 3\pi/2,\ \pm 5\pi/2,\ \pm 7\pi/2,\ ....$$

where $\sin^2\Phi = 1$ is a maximum. With

$$\Phi = 3\pi/2 = 4.712 = \pi D\sin\theta/\lambda$$

$$D = 4.712(441.16)/\pi\cdot\sin 5^\circ = 7.6 \times 10^{-6}\ \text{m}$$

and $D/\lambda \simeq 17$. An <u>exact</u> analysis locates the maxima by setting

$$d[\sin\Phi/\Phi]/d\Phi = 0$$

Thus,

$$d[\sin\Phi/\Phi]/d\Phi = \cos\Phi/\Phi - \sin\Phi/\Phi^2 = 0$$

This gives

$$\tan\Phi = \Phi$$

This is a transcendental equation. The solution $\Phi = 0$ correponds to the central peak ($\theta = 0$). The next solution is $\Phi = 4.493$. Solving

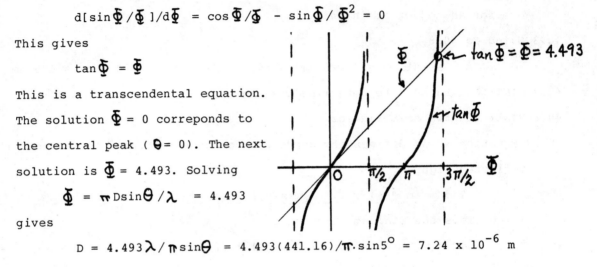

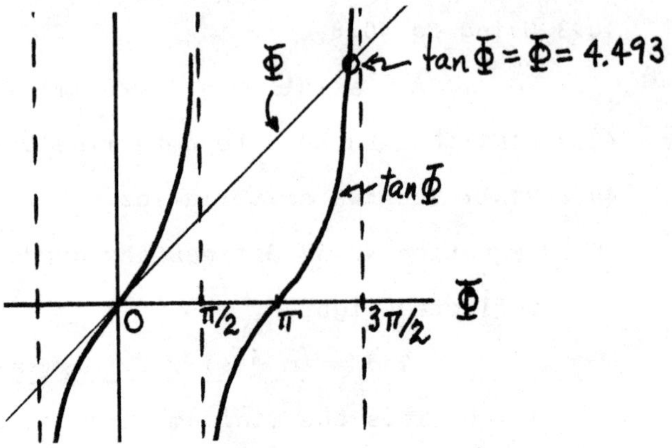

$$\Phi = \pi D\sin\theta/\lambda = 4.493$$

gives

$$D = 4.493\lambda/\pi\sin\theta = 4.493(441.16)/\pi\cdot\sin 5^{\circ} = 7.24 \times 10^{-6} \text{ m}$$

and $D/\lambda = 16.4$.

41.9 HINT: Use the circular-aperture relation $\sin\theta = 1.22(\lambda/D)$.

41.11 HINT: The objects are resolved if their angular separation is greater than the angular resolution (see Eq 41.24) of the telescope.

41.20 The resolving power is given by Eq 41.44,

$$R = mN$$

The number of lines equals the width of the grating divided by the line spacing

$$N = \text{width/spacing} = 0.1062 \text{ m}/2.313 \times 10^{-6} \text{ m} = 45{,}910$$

For the first-order spectrum m = 1 and

$$R = N = 45{,}910$$

42.4 We can use Eq 42.3,

$$E = h\nu$$

and the kinematic relation for waves

$$c = \nu\lambda$$

to obtain

$$\lambda = hc/E = (6.63 \times 10^{-34} \text{ J}\cdot\text{s})(3 \times 10^{8} \text{ m/s})/(1 \text{ eV})(1.60 \times 10^{-19} \text{ J/eV})$$

$$= 1.24 \times 10^{-6} \text{ m} = 1240 \text{ nm}$$

42.6 HINT: Make use of energy conservation, as embodied in Eq 42.15.

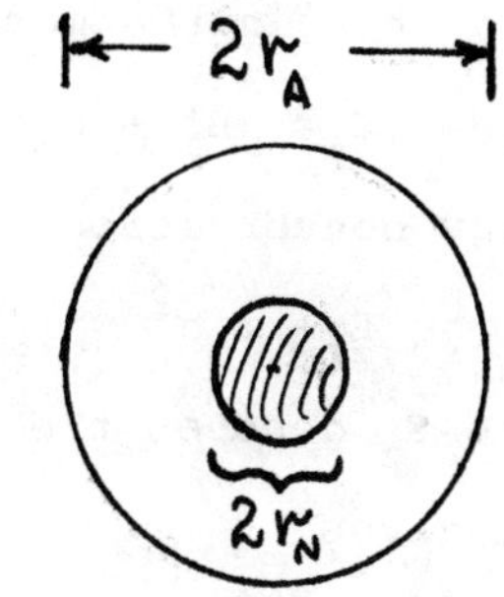

42.7 HINT: Imagine the atom as a sphere of radius r_A and the nucleus as a sphere of radius r_N. The fraction of alpha particles that suffer head-on collisions with the nucleus is approximately the fraction of the atomic cross-sectional area that is "blocked" by the nucleus.

42.9 HINT: The overall result is to transform kinetic energy into photon energy, with energy being conserved.

42.14 Using Eq 42.28,

$$\lambda = h/p = h/[2mK]^{\frac{1}{2}}$$

gives for the kinetic energy K,

$$K = (h/\lambda)^2/2m$$
$$= (6.63 \times 10^{-34}/1.65 \times 10^{-10})^2/2(0.911 \times 10^{-30})$$
$$= 8.86 \times 10^{-18}\ \cancel{J}\ (1\ \text{eV}/1.60 \times 10^{-19}\ \cancel{J})$$
$$= 55.4\ \text{eV}$$

The applied potential difference supplied 54.0 eV. The crystal supplied the remaining 1.4 eV.

43.1 a) Consider a head-on collision in which the photon recoils backward. The diagram shows the notation used to describe the photon and proton before and after the collision. Conservation of linear momentum is expressed as

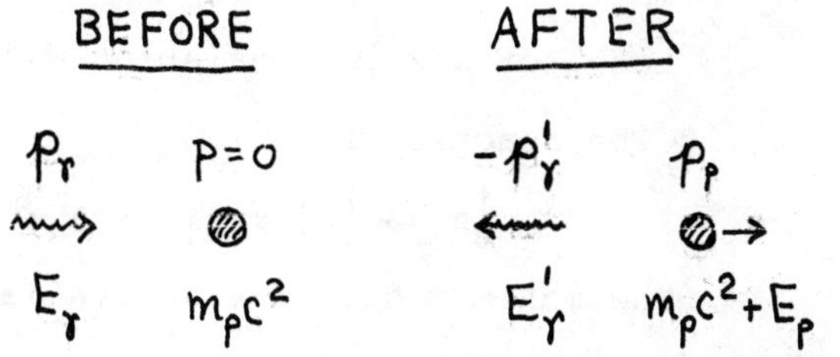

$$p_{before} = p_{after}$$

In terms of the notation in the diagram,

$$p_\gamma + 0 = -p'_\gamma + p_p$$

The minus sign accounts for the fact that the recoiling photon is traveling in the direction opposite to the incident photon. The energy and linear momentum for a photon are related by

$$p_{photon} = E_{photon}/c$$

Using this in the momentum conservation equation gives

$$E_\gamma = -E'_\gamma + p_p c \qquad (1)$$

Energy conservation is expressed by

$$E_\gamma + m_p c^2 = E'_\gamma + m_p c^2 + E_p$$

where E_p denotes the kinetic energy of the proton. Canceling the common term gives

$$E_\gamma = E'_\gamma + E_p \qquad (2)$$

Adding Eqs (1) and (2) eliminates E'_γ and gives

$$E_\gamma = \tfrac{1}{2}(E_p + p_p c) \qquad (3)$$

The proton momentum is related to the proton kinetic energy by

$$p_p^2/2m = E_p$$

so

$$\tfrac{1}{2}p_p c = [m_p c^2 E_p/2]^{\frac{1}{2}}$$

Using this in Eq (3) gives the desired result

$$E = \tfrac{1}{2}E_p + [m_p c^2 E_p/2]^{\frac{1}{2}}$$

b) With $E_p = 6$ MeV and $m_p c^2 = 938$ MeV

$$E = \tfrac{1}{2}(6) + [(938)(6)/2]^{\frac{1}{2}} = 56 \text{ MeV}$$

43.7 If 1,2,3 and 4 denote the protons then the combinations

1,2	1,3	1,4
	2,3	2,4
		3,4

enumerate all possible pairings - a total of six pairings. This total of six is consistent with $Z(Z-1)/2 = 4(3)/2 = 6$ for $Z = 4$.

43.10 The reaction is

$${}_0n^1 \rightarrow {}_1H^1 + {}_{-1}e + \nu$$

The maximum beta energy available is the difference in rest mass energy between the neutron and the decay products. Noting that the mass of the proton plus electron equals the hydrogen atom mass (ignoring the atomic binding energy) we have

$$E_\beta(\max) = (m_n - m_p - m_e)c^2 = (m_n - m_H)c^2$$

$$= (1.008665 - 1.007825)u \cdot (931.48 \text{ MeV/u}) = 0.782 \text{ MeV}$$

43.13 HINT: You should find that the difference between the rest mass of the Pu and the decay products is 0.000563u.

43.14 HINT: The time dependence of the activity is given by Eq 43.21.

43.19 The number of parent nuclei is given by

$$n_p = n_o e^{-\lambda_p t}$$

The parent decay rate is

$$dn_p/dt = -\lambda_p n_o e^{-\lambda_p t} = -\lambda_p n_p$$

The rate of parent decay equals the rate of daughter production. The rate of daughter decay is given by $-\lambda_d n_d$. The net rate of change of daughter nuclei is

$$dn_d/dt = \text{rate of production} - \text{rate of decay}$$

$$= \lambda_p n_p - \lambda_d n_d$$

When the net rate of change of the daughter nuclei is zero, $dn_d/dt = 0$, and

$$\lambda_p n_p = \lambda_d n_d$$

43.20 HINT: Use Eq 43.30 and note that

$$I = 23.1/\text{hr}\cdot\text{gm} = 23.1/(60\ \text{min})\cdot\text{gm}$$

43.23 HINT: The decay rate per gram is given by Eq 43.26,

$$I_o = -\lambda N_o = -13.6/\text{min}\cdot\text{gm}$$

43.30 HINT: The excitation energy is given by

$$E_{excitation} = (M_{239} + M_n - M_{240})(931.48\ \text{MeV/u})$$

43.33 HINT: Fill in the missing entries

Generation	Total #	# Produced
0	2	
1	4	2 = 4-2
2	8	6 = 8-2
3	16	14 = 16-2
n	.....	......

43.36 HINT: 1 gram of ^{235}U is 1/235 of one mole.

43.40 HINT: One liter of water has a mass of 1000 grams. One mole of water has a mass of 18 grams.

The following group of computer programs allow you to explore certain problems in greater depth. Computer listings are given for three popular brands of microcomputers: Radio Shack, Apple, and IBM. The programs are written in the BASIC language.

The programs have NOT been optimized with respect to speed or sophistication. They purposely have been made relatively simple and transparent. You are encouraged to improve them where it suits your needs and interests. For example, none of the programs produces printed output, simply because not everyone has a printer available for their computer system. If you do have a printer a modest effort on your part will let you convert PRINT statements to LPRINT statements for the Radio Shack and IBM computers and to PR#1 statements for the Apple.

The directory which follows lists the programs and indicates points of reference in University Physics.

1. VELOCITY AND ACCELERATION. Problems 4.10, 4.18, 4.23.
2. BOUNCING BALLS AND INFINITE SERIES. Problem 4.39.
3. NUMERICAL INTEGRATION. Problem 7.15 (and many others in later chapters).
4. ESCAPE FROM EARTH. Chapter 8, Example 3 (p 161).
5. SCATTERING ANGLES. Problem 10.37.
6. ANGULAR MOMENTUM AND KEPLER'S SECOND LAW. Problem 11.6
7. DAMPED MOTION. Problems 14.3, 14.6, 14.8, 14.40, 14.41.
8. LINEAR SYSTEMS AND FOURIER SERIES. Sections 18.5 and 37.1.
9. DEEP WATER WAVES AND RIPPLES. Chapter 18 (Equation 18.21).
10. THE MAXWELLIAN DISTRIBUTION. Problems 25.22, 25.23, 25.25, 25.26, 25.27.
11. ELECTROSTATIC POTENTIAL AND E = 0 POSITIONS. Problems 26.36 and 27.11.
12. EXPLOITING SYMMETRY. Problems 26.36 and 27.11 revisited.
13. INFINITE NETWORKS AND ALGORITHMS. Problem 31.40.
14. THE BOHR ATOM. Problems 42.10, 42.11.

You can use your microcomputer to help establish the ideas of instantaneous velocity and instantaneous acceleration. The program VEL which follows lets you evaluate the average velocity for time intervals of your choice. The definition of average velocity

$$v_{av} = \text{displacement/elapsed time}$$

and the definition of instantaneous velocity

$$v = \lim_{\Delta t \to 0} \{x(t + \Delta t) - x(t)\}/\Delta t$$

show that the average velocity approaches the instantaneous velocity in the limit as the time interval Δt approaches zero.

The program VEL lets you define the functional relationship between the position variable x and the time t. For example, in Problem 4.10 the relationship is

$$x = -4.9t^2$$

You then input the initial time (t) and the time interval (Δt). The program then evaluates and displays the average velocity

$$v_{av} = \{x(t + \Delta t) - x(t)\}/\Delta t$$

By choosing successively smaller values of Δt you <u>may</u> be able to infer the instantaneous velocity. We say "<u>may</u>" because if you make Δt <u>too small</u> the computer will not be able to provide the needed precision.

The program ACC allows you to use the velocity-time relation to determine the average acceleration

$$a_{av} = \{v(t + \Delta t) - v(t)\}/\Delta t$$

The instantaneous acceleration follows as the limit

$$a = \lim_{\Delta t \to 0} \{v(t + \Delta t) - v(t)\}/\Delta t$$

You can use the VEL program with Problems 4.10 and 4.18. The ACC program can be used with Problem 4.23.

In both programs, there is an endless loop which lets you repeat the evaluation for a different time and/or time interval. After each pass through the loop the program displays a complete summary of all calculations. This makes it possible to use the programs to develop tables and graphs.

WARNING! In forming powers such as T^3, use T*T*T instead of T^3. If you use the special mathematical functions like SIN(T) and EXP(T) you limit yourself to single-precision accuracy.

```
10 REM ** RADIO SHACK AND IBM PC **
20 REM ** VEL **
30 GOTO 150
40 DIM A#(30), B#(30), S#(30)
50 CLS:DEFDBL A,B,D,F,S,T,V,X
60 DEF FNX(T) = 3.2*T-0.21*T*T*T
70 INPUT"INPUT INITIAL TIME";T:PRINT
80 INPUT"INPUT TIME INTERVAL";DT:PRINT
90 V = (FNX(T+DT/2) - FNX(T-DT/2))/DT
100 PRINT"INITIAL TIME = ";T;"  TIME INTERVAL = ";DT:PRINT
110 PRINT"AVERAGE VELOCITY = ";V:PRINT
120 S(K) = V: A(K) = T: B(K) = DT: K = K+1
130 PRINT"   VELOCITY                          TIME INTERVAL       TIME ":PRINT
140 FOR L = 0 TO K-1:PRINT S(L),B(L),A(L): NEXT L:PRINT:GOTO 70
150 CLS:PRINT"THIS PROGRAM LETS YOU EVALUATE THE AVERAGE VELOCITY":PRINT
160 PRINT"YOU MUST TYPE AND ENTER THE POSITION-TIME RELATION.":PRINT
170 FOR J = 1 TO 2000: NEXT J
180 PRINT"FOR EXAMPLE, IF THE POSITION-TIME RELATION IS":PRINT
190 PRINT"  X(T) = 30T":PRINT:FOR J = 1 TO 3000:NEXT J
200 PRINT"YOU WOULD TYPE":PRINT
210 PRINT" 60 DEF FNX(T) = 30*T":PRINT:FOR J = 1 TO 4000:NEXT J
220 PRINT"AFTER TYPING LINE 60, PRESS 'ENTER'":PRINT
230 FOR J = 1 TO 2000: NEXT J
240 PRINT"THEN TYPE 'GOTO 40' AND PRESS 'ENTER'":PRINT:END
```

```
10 REM ** RADIO SHACK AND IBM PC **
20 REM ** ACC **
30 GOTO 150
40 DIM C#(30), B#(30), S#(30)
50 CLS:DEFDBL A,B,D,F,S,T,V,C
60 DEF FNV(T) = 3.22*T*T
70 INPUT"INPUT INITIAL TIME";T:PRINT
80 INPUT"INPUT TIME INTERVAL";DT:PRINT
90 A = (FNV(T+DT/2) - FNV(T-DT/2))/DT
100 PRINT"INITIAL TIME = ";T;"  TIME INTERVAL = ";DT:PRINT
110 PRINT"AVERAGE ACCELERATION = ";A:PRINT
120 S(K) = A: C(K) = T: B(K) = DT: K = K+1
130 PRINT"   ACCELERATION                TIME INTERVAL       TIME ":PRINT
140 FOR L = 0 TO K-1:PRINT S(L),B(L),C(L): NEXT L:PRINT:GOTO 70
150 CLS:PRINT"THIS PROGRAM LETS YOU EVALUATE THE AVERAGE ACCELERATION.":PRINT
160 PRINT"YOU MUST TYPE AND ENTER THE VELOCITY-TIME RELATION.":PRINT
170 FOR J = 1 TO 2000: NEXT J
180 PRINT"FOR EXAMPLE, IF THE VELOCITY-TIME RELATION IS":PRINT
190 PRINT"  V(T) = 30T":PRINT:FOR J = 1 TO 3000:NEXT J
200 PRINT"YOU WOULD TYPE":PRINT
210 PRINT" 60 DEF FNV(T) = 30*T":PRINT:FOR J = 1 TO 4000:NEXT J
220 PRINT"AFTER TYPING LINE 60, PRESS 'ENTER'":PRINT
230 FOR J = 1 TO 2000: NEXT J
240 PRINT"THEN TYPE 'GOTO 40' AND PRESS 'ENTER'":PRINT:END
```

```
10  REM  ** APPLE **
20  REM  ** VEL **
30  GOTO 150
40  DIM A(30),B(30),S(30)
50  HOME
60  DEF  FN X(T) = 4.9 * T * T - 3.2 * T
70  INPUT "INPUT INITIAL TIME ";T: PRINT
80  INPUT "INPUT TIME INTERVAL ";DT: PRINT
90 V = ( FN X(T + DT / 2) -  FN X(T - DT / 2)) / DT
100  PRINT "INITIAL TIME = ";T;"  TIME INTERVAL = ";DT: PRINT
110  PRINT "AVERAGE VELOCITY = ";A: PRINT
120 S(K) = V:A(K) = T:B(K) = DT:K = K + 1
130  PRINT "  VELOCITY  TIME INTERVAL    TIME": PRINT
140  FOR L = 0 TO K - 1: PRINT S(L),B(L),A(L): NEXT L: PRINT : GOTO 70
150  HOME : PRINT "THIS PROGRAM LETS YOU EVALUATE": PRINT
160  PRINT "THE AVERAGE VELOCITY.": PRINT
170  PRINT "YOU MUST TYPE AND ENTER THE": PRINT
180  PRINT "POSITION-TIME RELATION.": PRINT : PRINT
190  FOR J = 1 TO 3500: NEXT J: PRINT : PRINT
200  PRINT "FOR EXAMPLE, IF THE POSITION IS": PRINT
210  PRINT "GIVEN BY   X(T) = 30T   YOU TYPE": PRINT
220  PRINT " 60 DEF FN X(T) = 30*T": PRINT : FOR J = 1 TO 4000: NEXT J
230  PRINT : PRINT "AFTER TYPING LINE 60 PRESS 'RETURN'": PRINT
240  FOR J = 1 TO 2000: NEXT J
250  PRINT "THEN TYPE 'GOTO 40' AND PRESS 'RETURN'": END
```

```
10  REM  ** APPLE **
20  REM  ** ACC **
30  GOTO 150
40  DIM C(30),B(30),S(30)
50  HOME
60  DEF  FN V(T) = 3 * T * T * T - 2.87 * T
70  INPUT "INPUT INITIAL TIME ";T: PRINT
80  INPUT "INPUT TIME INTERVAL ";DT: PRINT
90 A = ( FN V(T + DT / 2) -  FN V(T - DT / 2)) / DT
100  PRINT "INITIAL TIME = ";T;"  TIME INTERVAL = ";DT: PRINT
110  PRINT "AVERAGE ACCELERATION = ";A: PRINT
120 S(K) = A:C(K) = T:B(K) = DT:K = K + 1
130  PRINT "ACCELERATION  TIME INTERVAL    TIME": PRINT
140  FOR L = 0 TO K - 1: PRINT S(L),B(L),C(L): NEXT L: PRINT : GOTO 70
150  HOME : PRINT "THIS PROGRAM LETS YOU EVALUATE": PRINT
160  PRINT "THE AVERAGE ACCELERATION.": PRINT
170  PRINT "YOU MUST TYPE AND ENTER THE": PRINT
180  PRINT "VELOCITY-TIME RELATION.": PRINT : PRINT
190  FOR J = 1 TO 3500: NEXT J: PRINT : PRINT
200  PRINT "FOR EXAMPLE, IF THE VELOCITY IS": PRINT
210  PRINT "GIVEN BY  V(T) = 30T  YOU TYPE": PRINT
220  PRINT " 60 DEF FN V(T) = 30*T": PRINT : FOR J = 1 TO 4000: NEXT J
230  PRINT : PRINT "AFTER TYPING LINE 60 PRESS 'RETURN'": PRINT
240  FOR J = 1 TO 2000: NEXT J
250  PRINT "THEN TYPE 'GOTO 40' AND PRESS 'RETURN'": END
```

Problem 4.39 asks that you determine the distance covered by a ball which falls freely from a height of 4.9 m and then rebounds to a height that is 81% of the distance it fell. The sequence is repeated an infinite number of times. Each time the ball rebounds it rises to a height that is 81% of the distance it fell. As we will show, the solution of this problem results in an infinite series. We will show how this infinite series is evaluated. Then we will let you explore such series with your microcomputer.

The first fall to the floor gives a distance of 4.9 m. The ball rebounds to a height of (0.81)4.9 m and falls back an equal distance. When it strikes the floor for the second time it has traveled 4.9 + 2(0.81)4.9 m. After the second bounce it rises to a height of 0.81 times the distance it fell, or (0.81)(0.81)4.9 m = $(0.81)^2 4.9$ m. It falls back this same distance so that when it strikes the floor for the third time it has traveled $4.9 + 2(0.81)4.9 + 2(0.81)^2 4.9$ m. Each sucessive rise and fall covers a distance that is a factor of 0.81 times the distance covered on the preceding rise and fall. The total distance traveled at the moment the ball hits the floor for the Nth time can be expressed as the series

$$4.9 + 2(4.9)\sum_{k=1}^{N-1}(0.81)^k = D(N)$$

This series can be summed exactly whether N is finite or infinite, by recognizing that the infinite series

$$1 + x + x^2 + x^3 + \ldots\ldots = (1-x)^{-1} \qquad |x| < 1$$

Using this result the series D(N) can be summed. In the limit as N approaches infinity the series is

$$1 + 2x + 2x^2 + 2x^3 + \ldots\ldots = 2(1-x)^{-1} - 1 \qquad |x| < 1$$

With x = 0.81, the distance traveled in the limit as N approaches infinity is

$$4.9[2(1 - 0.81)^{-1} - 1] \text{ m} = 46.679 \text{ m} \qquad \text{(to 5 significant figures)}$$

However, because successive terms in the infinite series become progressively smaller, a finite number of terms will give the distance accurate to, say, 5 significant figures.

The program BOUNCE which follows allows you to evaluate D(N) for values of N which you choose. By choosing successively larger values of N you will see that the distance D(N) approaches the limiting value of 46.679 m. When you are finished the series of values of D(N) and N will be displayed for you.

The program also evaluates T(N), the time elapsed when the ball strikes the floor for the Nth time. You should convince yourself that 1 s elapses during the first fall to the floor and that 0.9 s elapses during its subsequent rise. The series for T(N) is essentially the same as the series for D(N), being given by

$$T(N) = 1 + 2(0.9) + 2(0.9)^2 + \ldots = 1 + 2\sum_{k=1}^{N-1}(0.9)^k$$

In the limit as N approaches infinity T(N) approaches

$$\lim_{N\to\infty} T(N) = 2[(1 - 0.9)^{-1} - 1]\ \text{s} = 19\ \text{s}$$

As you choose larger and larger values of N you will observe that T(N) approaches the limiting value of 19 s. When you are finished all of the values of T(N) are displayed along with the values of D(N) and N. Have fun.

Would you like to test your programming skills on an "infinite" series? The infinite series

$$S = \sum_{n=1}^{\infty} 1/n^4 = 1 + 1/16 + 1/81 + 1/256 + \ldots.$$

occurs in the theory of thermal radiation. This infinite series can be evaluated using a rather sophisticated Fourier series technique. The result is

$$S = \pi^4/90 = 1.08232\ldots.$$

Because successive terms become progressively smaller, the finite series

$$S(N) = \sum_{n=1}^{N} 1/n^4$$

should be a rather good approximation to S.

Write a short program that evaluates S(N) for N = 10, 50, 100, etc. You may conclude that "infinity" is not so large after all.

```
10 REM ** RADIO SHACK AND IBM PC **
20 REM ** BOUNCE **
30 DIM X(100), Y(100),Z(100):CLS:PRINT:PRINT
40 PRINT TAB(5) "THIS PROGRAM LETS YOU SOLVE 4.39":PRINT
50 PRINT TAB(5) "BY SUMMING THE DISTANCES AND TIMES FOR A":PRINT
60 PRINT TAB(5) "FINITE NUMBER OF BOUNCES. THE DISTANCE":PRINT
70 PRINT TAB(5)"AND TIME APPROACH FINITE LIMITS.":PRINT:PRINT
80 FOR J = 1 TO 4000: NEXT J
90 CLS:J = 1:D = 4.9: S = 9.8:U = 2:T = 1: PRINT
100 PRINT TAB(10) "START WITH A VALUE OF N BETWEEN 15 AND 25.":PRINT
110 PRINT TAB(10) "TYPE THE VALUE OF N AND DEPRESS 'ENTER'.":PRINT
120 INPUT N
130 FOR K = 1 TO N-1: D = D+0.81*S: T = T+0.9*U: S = 0.81*S: U = 0.9*U:NEXT K
140 X(J) = D: Y(J) = N: Z(J) = T :CLS
150 PRINT TAB(14) "D =";D;"M      T =";T;"S      N =";N:PRINT:PRINT
160 PRINT TAB(10) "TO REPEAT FOR ANOTHER VALUE OF N,":PRINT
170 PRINT TAB(10) "TYPE R AND DEPRESS 'ENTER'.":PRINT
180 PRINT TAB(10) "TO QUIT, TYPE Q AND DEPRESS 'ENTER'.":PRINT
190 INPUT A$:IF A$ = "Q" THEN 230
200 J = J+1:D = 4.9: S = 9.8:T=1:U=2:PRINT:CLS:PRINT:PRINT:PRINT
210 PRINT TAB(20) "INPUT A VALUE OF N":PRINT:INPUT N:CLS
220 GOTO 130
230 CLS:PRINT" DISTANCE<M>       TIME<S>         N":PRINT:FOR L = 1 TO J
240 PRINT TAB(2)X(L),Z(L),Y(L):NEXT L
250 PRINT:PRINT"TO RUN AGAIN, TYPE 'GO' AND PRESS 'ENTER'":PRINT
260 PRINT"TO QUIT, TYPE 'Q' AND PRESS 'ENTER'":PRINT
270 INPUT S$: IF S$= "GO" THEN 90ELSE END
```

```
10 REM ** APPLE **
20 REM ** BOUNCE **
30 DIM X(100),Y(100),Z(100): HOME : PRINT : PRINT
40 PRINT TAB( 3)"THIS PROGRAM LETS YOU SOLVE 4.39 BY": PRINT
50 PRINT TAB( 3)"SUMMING THE DISTANCES AND TIMES FOR A": PRINT
60 PRINT TAB( 3)"FINITE NUMBER OF BOUNCES. THE DISTANCE"
70 PRINT TAB( 3)"AND TIME APPROACH FINITE LIMITS.": PRINT
80 FOR J = 1 TO 6000: NEXT J
90 HOME :J = 1:D = 4.9:S = 9.8:U = 2:T = 1: PRINT : PRINT
100 PRINT TAB( 5)"START WITH A VALUE OF": PRINT
110 PRINT TAB( 5)"N BETWEEN 15 AND 25.": PRINT
120 PRINT TAB( 5)"TYPE N AND PRESS 'RETURN'": PRINT : INPUT N
130 FOR K = 1 TO N - 1:D = D + 0.81 * S:T = T + 0.9 * U:S = 0.81 * S
140 U = 0.9 * U: NEXT K:X(J) = D:Y(J) = N:Z(J) = T: HOME
150 PRINT TAB( 5)"D = "; INT (1E4 * (D + .00005)) / 1E4;" M": PRINT
160 PRINT TAB( 5)"T = "; INT (1E4 * (T + .00005)) / 1E4;" S   N = ";N: PRINT
170 PRINT "TO REPEAT FOR ANOTHER VALUE OF N": PRINT
180 PRINT "TYPE 'R' AND PRESS 'RETURN'": PRINT
190 PRINT "TO QUIT, TYPE 'Q' AND PRESS 'RETURN'.": PRINT
200 INPUT A$: IF A$ = "Q" THEN 250
210 J = J + 1:D = 4.9:S = 9.8:T = 1:U = 2
220 HOME : PRINT : PRINT : PRINT : PRINT
230 PRINT TAB( 5)"INPUT A VALUE OF N": PRINT : INPUT N
240 HOME : GOTO 130
250 HOME : PRINT "  DISTANCE<M>  TIME<S>          N": PRINT : FOR L = 1 TO J
260 X(L) = INT (1E4 * (X(L) + .00005)) / 1E4
270 Z(L) = INT (1E4 * (Z(L) + .00005)) / 1E4
280 PRINT TAB( 5)X(L),Z(L),Y(L): NEXT L
290 PRINT : PRINT "TO REPEAT, TYPE 'GO' AND PRESS 'RETURN'.": PRINT
300 PRINT "TO QUIT, TYPE 'Q' AND PRESS 'RETURN'": PRINT
310 INPUT S$: IF S$ = "GO" THEN 90ELSE END
```

There are many instances where you will find it necessary to perform an integration. For example, in Problem 7.15 you are asked to determine the work done by a force by evaluating the integral

$$\int_{r_1}^{r_2} F dr = F_o \int_{r_1}^{r_2} [2(\sigma/r)^{13} - (\sigma/r)^7] dr$$

Many such integrals can be evaluated analytically, including the one above. In situations where you are unable to perform the integration analytically, a numerical integration technique can be used. The numerical integration program allows you to choose the integrand and the lower limit and the upper limit.

The program evaluates the integral using Simpson's rule. Many calculus textbooks have a discussion of Simpson's rule.*

Although the program gives step-by-step instructions on how to proceed, the following precautions and recommendations should be kept in mind:

1) Convert the integration variable to dimensionless form. In the integral above, the integration variable is r, a length. A suitable dimensionless variable is the ratio r/σ. Thus, with

$$u = r/\sigma \quad ; \; dr = \sigma du$$

the integral becomes

$$\int_{r_1}^{r_2} F dr = \sigma F_o \int_{u_1}^{u_2} [2u^{-13} - u^{-7}] du \quad ; \; u_1 = r_1/\sigma \quad , \; u_2 = r_2/\sigma$$

By using dimensionless variables you free the computer from dealing with the specific values of certain dimensional quantities, like σ in the integral above.

2) You must avoid division by zero. For example, if you try to use the program to evaluate the integral

$$I = \int_0^{\infty} x^3 dx/(e^x - 1)$$

you will get an error message indicating division by zero. The reason is that at the lower limit, x = 0, the program sets x = 0 and finds that

$$x^3/(e^x-1) = 0^3/(e^0-1) = 0/(1-1) = 0/0$$

* See, for example, Calculus with analytic geometry. E W Swokowski. Prindle, Weber & Schmidt. Boston 1983

In fact, the limit as x approaches zero of $x^3/(e^x-1)$ is zero, but the computer program does not "sense" this - it is asked to divide zero by zero and wisely refuses. One way to handle this type of situation is to set the lower limit equal to, say, 0.001, and then repeat the integration for a lower limit of 0.0001. If you obtain essentially equal results with different lower limits you can safely conclude that the small range of the integration from 0 to 0.0001 or 0 to 0.001 makes an ignorable contribution to the integral.

3) What about integrals with large - even infinite - ranges? The integral

$$I = \int_0^\infty x^3 dx/(e^x-1)$$

is again a good example. In this integral and many similar ones the integration can be terminated at a finite upper limit without significant error. For example, you can set the upper limit equal to 10, and then repeat the integration with an upper limit of 20. Comparing results will let you decide when the upper limit is large enough.

4) Simpson's rule divides the range of integration into a number of segments. The time required to perform the numerical integration increases as the number of segments increases. The precision of the method also increases as the number of segments increases. How can you decide whether or not you have chosen a large enough number of segments? Start by choosing a modest number of segments and then repeat with a larger number of segments. Repeat, with a larger and larger number of segments until the value of the integral does not change significantly. The program summarizes all such repetitions.

5) To convince yourself that the program is running properly you may want to evaluate an integral for which you know the exact value. You are urged to get some practice by using the program to evaluate

$$I_1 = \int_0^\infty e^{-x} dx = 1 \qquad ; \qquad I_2 = \int_0^{2\pi} \cos^2 x \cdot dx = \pi = 3.14159$$

$$I_3 = \int_0^\infty x^3 dx/(e^x-1) = \pi^4/15 = 6.49394$$

```
10 REM ** RADIO SHACK **
20 REM ** INTEGRAL **
30 CLS
40 PRINT"THIS IS A NUMERICAL INTEGRATION PROGRAM"
50 PRINT:PRINT"IT USES SIMPSON'S RULE"
60 PRINT:PRINT"YOU GET TO CHOOSE THE INTEGRAND IN A MOMENT":PRINT
70 GOTO 260
80 CLS:PRINT:INPUT"INPUT LOWER LIMIT ";A:PRINT
90 INPUT"INPUT UPPER LIMIT ";B:PRINT
100 PRINT"INPUT THE NUMBER OF SEGMENTS":PRINT
110 INPUT"CHOOSE AN EVEN NUMBER ";N:PRINT
120 H = (B-A)/N
130 DEF FNF(X) = EXP(-X)
140 S = H*(FNF(A) + 4*FNF(A+H) + FNF(B))/3
150 FOR J = 2 TO N-2 STEP 2: X = A+J*H: S = S+2*H*(FNF(X)+2*FNF(X+H))/3
160 NEXT J:PRINT:PRINT"LOWER LIMIT = ";A;"  UPPER LIMIT = ";B;"   # SEGMENTS = ";N
170 PRINT:PRINT"        INTEGRAL = ";S:PRINT:K = K+1
180 PRINT:D(K)=S:L(K)=A:U(K)=B:R(K)=N
190 PRINT:PRINT"TO RUN AGAIN WITH SAME INTEGRAND TYPE 'R'":PRINT
200 PRINT"TO QUIT TYPE 'Q'":PRINT
210 A$ = INKEY$
220 IF A$= ""THEN 210ELSE 230
230 IF A$= "Q" THEN 240ELSE 430
240 CLS:PRINT TAB(3)"SUMMARY OF CALCULATIONS (N DENOTES # OF SEGMENTS)"
250 PRINT:GOTO 390
260 PRINT"WHEN THE MONITOR READS 'READY', TYPE THE FOLLOWING:"
270 FOR J = 1 TO 2000: NEXT J
280 PRINT: PRINT"130 DEF FNF(X) = 'YOUR CHOICE OF INTEGRAND'"
```

CONTINUES ON NEXT PAGE

```
290 FOR J = 1 TO 2500: NEXT J:PRINT
300 PRINT"FOR EXAMPLE, TO GET X AS THE INTEGRAND YOU WOULD TYPE":PRINT
310 PRINT"130 DEF FNF(X) = X":PRINT
320 FOR J = 1 TO 2000: NEXT J:PRINT
330 PRINT"TO GET E(-X) AS THE INTEGRAND YOU WOULD TYPE":PRINT
340 PRINT"130 DEF FNF(X) = EXP(-X)":PRINT
350 FOR J = 1 TO 2000: NEXT J
360 PRINT:PRINT"AFTER TYPING LINE 130 PRESS 'ENTER'":PRINT
370 FOR J = 1 TO 500: NEXT J
380 PRINT"THEN TYPE 'GOTO 80' AND PRESS 'ENTER'":PRINT:END
390 FOR J = 1 TO K
400 PRINT"INTEGRAL = ";D(J);"  N = ";R(J):PRINT
410 PRINT"LOWER LIMIT = ";L(J);"  UPPER LIMIT = ";U(J):PRINT:PRINT
420 NEXT J:END
430 S=0:GOTO 80
```

```
10 REM ** IBM PC **
20 REM ** INTEGRAL **
30 CLS
40 PRINT"THIS IS A NUMERICAL INTEGRATION PROGRAM"
50 PRINT:PRINT"IT USES SIMPSON'S RULE"
60 PRINT:PRINT"YOU GET TO CHOOSE THE INTEGRAND IN A MOMENT":PRINT
70 GOTO 260
80 CLS:PRINT:INPUT"INPUT LOWER LIMIT ";A:PRINT
90 INPUT"INPUT UPPER LIMIT ";B:PRINT
100 PRINT"INPUT THE NUMBER OF SEGMENTS":PRINT
110 INPUT"CHOOSE AN EVEN NUMBER ";N:PRINT
120 H = (B-A)/N
```

CONTINUES ON NEXT PAGE

```
130 DEF FNF(X) = EXP(-X)
140 S = H*(FNF(A) + 4*FNF(A+H) + FNF(B))/3
150 FOR J = 2 TO N-2 STEP 2: X = A+J*H: S = S+2*H*(FNF(X)+2*FNF(X+H))/3
160 NEXT J:PRINT:PRINT"LOWER LIMIT = ";A;"  UPPER LIMIT = ";B;"   # SEGMENTS = ";N
170 PRINT:PRINT"        INTEGRAL = ";S:PRINT:K = K+1
180 PRINT:D(K)=S:L(K)=A:U(K)=B:R(K)=N
190 PRINT:PRINT"TO RUN AGAIN WITH SAME INTEGRAND TYPE 'R'":PRINT
200 PRINT"TO QUIT TYPE 'Q'":PRINT
210 A$ = INKEY$
220 IF A$= ""THEN 210ELSE 230
230 IF A$= "Q" THEN 240ELSE 430
240 CLS:PRINT TAB(3)"SUMMARY OF CALCULATIONS (N DENOTES # OF SEGMENTS)"
250 PRINT:GOTO 390
260 PRINT"WHEN THE MONITOR READS 'GO', TYPE THE FOLLOWING:"
270 FOR J = 1 TO 2000: NEXT J
280 PRINT: PRINT"130 DEF FNF(X) = 'YOUR CHOICE OF INTEGRAND'"
290 FOR J = 1 TO 2500: NEXT J:PRINT
300 PRINT"FOR EXAMPLE, TO GET X AS THE INTEGRAND YOU WOULD TYPE":PRINT
310 PRINT"130 DEF FNF(X) = X":PRINT
320 FOR J = 1 TO 2000: NEXT J:PRINT
330 PRINT"TO GET E(-X) AS THE INTEGRAND YOU WOULD TYPE":PRINT
340 PRINT"130 DEF FNF(X) = EXP(-X)":PRINT
350 FOR J = 1 TO 2000: NEXT J
360 PRINT:PRINT"AFTER TYPING LINE 130 PRESS 'ENTER'":PRINT
370 FOR J = 1 TO 500: NEXT J
380 PRINT"THEN TYPE 'GOTO 80' AND PRESS 'ENTER'":PRINT:END
390 FOR J = 1 TO K
400 PRINT"INTEGRAL = ";D(J);"  N = ";R(J):PRINT
410 PRINT"LOWER LIMIT = ";L(J);"  UPPER LIMIT = ";U(J):PRINT:PRINT
420 NEXT J:END
430 S=0:GOTO 80
```

```
10  REM  ** APPLE **
20  REM  ** INTEGRAL **
30  HOME : PRINT "THIS IS A NUMERICAL INTEGRATION PROGRAM": PRINT
40  PRINT "IT USES SIMPSON'S RULE": PRINT
50  PRINT "YOU GET TO CHOOSE THE INTEGRAND": PRINT
60  GOTO 270
70  HOME : PRINT : PRINT "INPUT LOWER LIMIT": PRINT : INPUT A
80  PRINT : PRINT "INPUT UPPER LIMIT": PRINT : INPUT B
90  PRINT : PRINT "INPUT NUMBER OF SEGMENTS": PRINT
100  PRINT "CHOOSE AN EVEN NUMBER": PRINT : INPUT N:H = (B - A) / N: HOME
110  DEF  FN F(X) =  EXP ( - X)
120 S = H * ( FN F(A) + 4 *  FN F(A + H) +  FN F(B)) / 3
130  FOR J = 2 TO N - 2 STEP 2
140 X = A + J * H
150 S = S + 2 * H * ( FN F(X) + 2 *  FN F(X + H)) / 3
160  NEXT J
170  PRINT : PRINT "LOWER LIMIT = ";A;"  UPPER LIMIT = ";B
180  PRINT : PRINT "NUMBER OF SEGMENTS = ";N: PRINT
190  PRINT : PRINT "    INTEGRAL = ";S: PRINT :K = K + 1
200  PRINT :D(K) = S:L(K) = A:U(K) = B:R(K) = N
210  PRINT "TO REPEAT WITH SAME INTEGRAND ENTER 'R'": PRINT
220  PRINT "TO QUIT ENTER 'Q'": PRINT
230  INPUT A$
240  IF A$ = "R" THEN 470
250  HOME : PRINT "SUMMARY (N = # SEGMENTS)"
260  PRINT : GOTO 430
270  FOR J = 1 TO 3000: NEXT J: PRINT "WHEN THE CURSOR BLINKS, TYPE": PRINT
280  FOR J = 1 TO 3000: NEXT J
```

CONTINUES ON NEXT PAGE

```
290 PRINT : PRINT " 110 DEF FN F(X) = 'YOUR INTEGRAND'"
300 FOR J = 1 TO 3500: NEXT J: PRINT
310 PRINT "FOR EXAMPLE, TO GET X AS THE INTEGRAND": PRINT
320 PRINT "YOU WOULD TYPE": PRINT
330 PRINT " 110 DEF FN F(X) = X": PRINT
340 FOR J = 1 TO 3000: NEXT J
350 PRINT "TO GET EXP(-X) AS THE INTEGRAND TYPE": PRINT
360 PRINT " 110 DEF FN F(X) = EXP(-X)": PRINT
370 FOR J = 1 TO 3000: NEXT J
380 PRINT : PRINT "AFTER TYPING LINE 110 PRESS 'RETURN'": PRINT : PRINT
390 FOR J = 1 TO 1500: NEXT J
400 PRINT "THEN TYPE 'GOTO 70' AND PRESS 'RETURN'": END
410 PRINT
420 END
430 FOR J = 1 TO K
440 PRINT "INTEGRAL = ";D(J);"  N = ";R(J): PRINT
450 PRINT "LOWER LIMIT = ";L(J);"  UPPER LIMIT = ";U(J): PRINT : PRINT
460 NEXT J: END
470 S = 0: GOTO 70
```

In Example 3 of Chapter 8 (p 161) the conservation of mechanical energy is used to determine the escape speed of a projectile launched from the surface of the earth. The minimum escape speed is shown to be

$$v_{escape} = [2GM_E/R_E]^{\frac{1}{2}} = 11.2 \text{ km/s}$$

A projectile launched radially outward from the surface of the earth with this speed would not fall back to the surface. The analysis leading to this escape speed ignores air friction, the earth's rotation, and the gravitational action of the moon and sun.

The program FAROUT enables you to follow the motion of a projectile launched from the surface of the earth. You input the initial speed of the projectile. You also input an "escape distance" - a distance beyond which you presume the earth's gravitational force is smaller than the ignored gravity of the moon, sun, etc.

The program integrates Newton's second law to calculate the velocity and position at 2 s intervals. The position is displayed at 5 min intervals. If the projectile velocity becomes negative - signalling that it is falling back toward the surface - then the program halts and informs you of the conditions. If the projectile reaches your imposed escape distance then the program also halts and informs you of the results.

You can study the escape from other objects (our moon, or an asteroid) by changing the mass (M) and radius (R) that appear as constants in statement number 50.

```
10 REM ** RADIO SHACK AND IBM PC **
20 REM ** FAROUT **
30 S = 1: K = 1/2
40 CLS
50 G = 6.6732E-11: M = 5.98E24: R = 6.38E6
60 A = G*M/(R*R)
70 INPUT"INPUT ESCAPE DISTANCE, AS MULTIPLE OF EARTH RADIUS ";N
80 PRINT:INPUT"INPUT INITIAL VELOCITY IN KM/S ";V
90 V0 = V
100 V = V*1000
110 CLS:PRINT
120 PRINT" TIME [MIN]   VELOCITY [KM/S]  DISTANCE [R(EARTH)]":PRINT
130 T=0
140 GOTO220
150 FOR J = 1 TO 150
160 V = V - K*2*A/(S*S)
170 IF V < 0 THEN 240
180 S = S + 2*V/R: K = 1
190 IF S > N THEN 270
200 NEXT J
210 T = T + 5
220 PRINT"     "; T,INT(V)/1000,INT(1000*S)/1000
230 GOTO 150
240 PRINT:PRINT"PROJECTILE REACHED MAXIMUM ALTITUDE OF ";S;" R(EARTH)":PRINT
250 PRINT"INITIAL VELOCITY = ";V0;" KM/S":PRINT
260 PRINT"TO RUN AGAIN, ENTER 'RUN'":PRINT:END
270 PRINT:PRINT"PROJECTILE REACHED ESCAPE DISTANCE OF ";N;" TIMES EARTH RADIUS"
280 PRINT:PRINT"INITIAL VELOCITY = ";V0;" KM/S":PRINT
290 PRINT"TO RUN AGAIN, ENTER 'RUN'":PRINT:END
```

```
10  REM  ** APPLE **
20  REM  ** FAROUT **
30 S = 1:K = 1 / 2: HOME
40 G = 6.6732E - 11:M = 5.98E24:R = 6.38E6
50 A = G * M / (R * R)
60  PRINT "INPUT ESCAPE DISTANCE AS": PRINT
70  INPUT "A MULTIPLE OF EARTH RADIUS ";N: PRINT
80  INPUT "INPUT INITIAL VELOCITY IN KM/S ";V: PRINT
90 V0 = V:V = 1000 * V: HOME : PRINT
100  PRINT "TIME(MIN) VELOCITY(KM/S) DISTANCE(RE)": PRINT
110 T = 0
120  GOTO 190
130  FOR J = 1 TO 150
140 V = V - K * 2 * A / (S * S)
150  IF V < 0 THEN 210
160 S = S + 2 * V / R:K = 1
170  IF S > N THEN 240
180  NEXT J:T = T + 5
190  PRINT T, INT (V) / 1000, INT (1000 * S) / 1000
200  GOTO 130
210  PRINT : PRINT "PROJECTILE REACHED R = ";S;" RE": PRINT
220  PRINT "INITIAL VELOCITY = ";V0;" KM/S": PRINT
230  PRINT "TO RUN AGAIN, ENTER 'RUN'": PRINT : END
240  PRINT : PRINT "PROJECTILE REACHED R = ";N;" RE": PRINT
250  PRINT "INITIAL VELOCITY = ";V0;" KM/S": PRINT
260  PRINT "TO RUN AGAIN, ENTER 'RUN'": END
```

The figure (similar to Fig 10.28, p 203) shows how the angle of scatter (θ) in the lab frame is related to the angle of scatter (θ') in the center of mass frame. Equation 10.44 enables you to determine the lab angle θ by plugging in the center of mass angle, θ'.

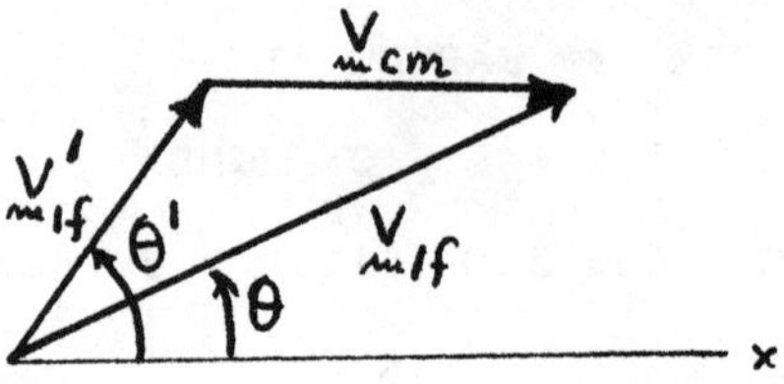

$\tan\theta = \sin\theta'/[\cos\theta' + (m_{inc}/m_{target})]$

The inverse problem - finding the center of mass angle for a given lab angle - is illustrated in Example 10 (p 203) for the special case $\theta = 90^{o}$.

The program CM allows you to evaluate the center of mass angle of scatter for the full range of lab angles. You input the mass of the incident particle and the mass of the target particle. Because the relation between the scattering angles involves only the ratio of these two masses you need not input actual masses - it is sufficient to use relative masses. For example, to describe the scatter of a helium atom (atomic weight 4) incident on a target carbon atom (atomic weight 12) you simply input masses of 4 and 12.

When the mass of the target particle is less than or equal to the mass of the incident particle [picture: a bowling ball smashes into a stationary bb] there is a maximum scattering angle in the lab frame given by

$\theta_{max} = \sin^{-1}(m_{target}/m_{inc})$

After determining the range of allowed lab angle of scatter, the program CM uses the inverse of Eq 10.44 to evaluate the center of mass angle. Corresponding values of the lab and center of mass angles are displayed for the full range of lab angle.

Where do you come in? It can be very instructive to sketch the velocity vectors ($\mathbf{v}'_{1f}$, $\mathbf{v}_{1f}$, and $\mathbf{v}_{cm}$ in the figure above) for several different pairs of the scattering angles. Try running the program for three cases:

$m_{inc}/m_{target} << 1$, $= 1$, and $>> 1$

For example, an alpha particle incident on a ^{238}U nucleus gives $m_{inc}/m_{target} = 4/238 << 1$.

Problem 10.37 can be tackled with the CM program. The incident alpha particle scattered by a cobalt nucleus gives a mass ratio $m_{inc}/m_{target} = 4/60$.

```
10 REM ** RADIO SHACK AND IBM PC **
20 REM ** CM **
30 Q = 1:K = 3.14159/18
40 INPUT"INPUT INCIDENT MASS";M1:PRINT
50 INPUT"INPUT TARGET MASS";M2:PRINT:CLS
60 R = M1/M2
70 IF R => 1 THEN GOSUB 220
80 PRINT:PRINT TAB(2)"INCIDENT MASS/TARGET MASS = ";M1;"/";M2:PRINT
90 PRINT TAB(3)"LAB ANGLE     CM ANGLE":PRINT
100 FOR J = 0 TO 18
110 T = J*K
120 IF T < 1.5708 THEN 140
130 Q = -1
140 C = -R*SIN(T)^2 + Q*SQR( (R*SIN(T)^2)^2 - (R*SIN(T))^2 + COS(T)^2)
150 S = SQR(1-C*C): P = ATN(S/C)
170 IF S/C < 0 THEN P = P + 3.14159
180 IF C = -1 THEN P = 3.14159
190 PRINT TAB(4)  T*180/3.14159, (180/3.14159)*P
200 NEXT J
210 END
220 IF R = 1 THEN K = 3.14159/36
230 IF R > 1 THEN K = (1/18)*ATN(M2/SQR(M1^2-M2^2))
240 PRINT"WHEN MASS OF INCIDENT PARTICLE IS GREATER THAN OR EQUAL":PRINT
250 PRINT"TO MASS OF TARGET THERE IS A MAXIMUM SCATTERING ANGLE IN":PRINT
260 PRINT"THE LAB FRAME. THIS MAXIMUM ANGLE IS GIVEN BY THE EQUATION":PRINT
270 PRINT"SIN(MAX ANGLE) = TARGET MASS/INCIDENT MASS. FOR YOUR DATA,":PRINT
280 PRINT"MAXIMUM LAB ANGLE OF SCATTER = ";(180/3.14159)*18*K;" DEGREES"
290 FOR J = 1 TO 7000: NEXT J
300 PRINT:RETURN
```

```
10  REM  ** APPLE **
20  REM  ** CM **
30 K = 3.14159 / 18:Q = 1
40  INPUT "INPUT INCIDENT MASS";M1: PRINT
50  INPUT "INPUT TARGET MASS";M2: PRINT : HOME
60 R = M1 / M2: IF R =  > 1 THEN  GOSUB 190
70  PRINT : PRINT  TAB( 2)"INCIDENT MASS/TARGET MASS = ";M1;"/";M2: PRINT
80  PRINT  TAB( 3)"LAB ANGLE     CM ANGLE": PRINT
90  FOR J = 0 TO 18:T = J * K
100 Y = R * ( SIN (T)) ^ 2
110  IF T < 1.5708 THEN 130
120 Q =  - 1
130 C =  - Y + Q *  SQR (Y * Y - R * Y +  COS (T) ^ 2)
140 S =  SQR (1 - C * C):P =  ATN (S / C)
150  IF S / C < 0 THEN P = P + 3.14159
160  IF C =  - 1 THEN P = 3.14159
170  PRINT  TAB( 4)T * 180 / 3.14159,(180 / 3.14159) * P
180  NEXT J: END
190  IF R = 1 THEN K = 3.14159 / 36
200  IF R > 1 THEN K = (1 / 18) *  ATN (M2 /  SQR (M1 ^ 2 - M2 ^ 2))
210  PRINT "WHEN MASS OF INCIDENT PARTICLE IS": PRINT
220  PRINT "GREATER THAN OR EQUAL TO MASS OF": PRINT
230  PRINT "TARGET THERE IS A MAXIMUM SCATTERING": PRINT
240  PRINT "ANGLE IN THE LAB FRAME. THIS MAXIMUM": PRINT
250  PRINT "ANGLE IS GIVEN BY THE EQUATION": PRINT
260  PRINT "SIN(LAB MAX) = TARGET MASS/INCIDENT MASS": PRINT
270  PRINT "FOR YOUR DATA,": PRINT
280  PRINT "MAX LAB ANGLE = ";(180 / 3.14159) * 18 * K;" DEGREES"
290  FOR J = 1 TO 7000: NEXT J
300  PRINT : RETURN
```

It is shown in Chapter 11, Section 1, that angular momentum is conserved for a central force. Because the gravitational force acting between the sun and its satellites is a central force angular momentum is conserved. Kepler's second law of planetary motion is a consequence of the conservation of angular momentum:

A vector from the sun to a planet sweeps out equal areas in equal times.

The program KEPLER allows you to verify angular momentum conservation and Kepler's second law for satellites of the sun. You choose a satellite from the table (next page), and input the semi-major axis and eccentricity of its elliptical orbit. The program integrates Newton's second law to obtain the velocity and position of the satellite. The program stores the values of the position and velocity at intervals of one-twelfth of the orbital period of the satellite. Using the computed position and velocity values the program then evaluates the angular momentum of the satellite at the end of each interval. The area swept out during each interval is calculated and displayed. You are given the option of displaying the position at intervals of one-twelfth of the orbital period.

The table on the next page gives the eccentricity of the orbit. For a circular orbit the eccentricity is zero and the satellite moves in uniform circular motion. In such a case the conservation of angular momentum and Kepler's second law are rather obvious. The larger the eccentricity the more the orbit deviates from a circular shape, and the greater the variation in speed along the orbit.

The satellites include major planets, minor planets (or asteroids), and comets. The computer displays the area swept out in units of square meters. The quantity displayed under the angular momentum heading is actually a pure number that is proportional to the angular momentum. This is made necessary by the fact that there are no reliable measurements of mass for the comets and the asteroids.

Problem 11.6 deals with the rate at which the earth-sun radius sweeps out area per second. You may want to test the program by using it to evaluate the area swept out per month and then convert the computer result to area swept out per second.

Satellite	Semi-Major Axis (AU)	Eccentricity
Mercury	0.3871	0.2056
Venus	0.7233	0.0068
Earth	1	0.0167
Mars	1.5237	0.0934
Jupiter	5.2028	0.0485
Saturn	9.5388	0.0557
Minor Planets		
Apollo	1.486	0.566
Eros	1.458	0.223
Psyche	2.923	0.135
Adonis	1.969	0.779
Icarus	1.078	0.827
Comets		
Whipple	3.80	0.35
Wolf	4.15	0.40
Temple	3.0	0.55
Oterma	3.96	0.14

```
10 REM ** RADIO SHACK AND IBM PC **
20 REM ** KEPLER **
30 DIM B(15), Q(15),X(15),Y(15),L(15)
40 INPUT"INPUT SEMI-MAJOR AXIS IN AU";A
50 PRINT:INPUT"INPUT ECCENTRICITY";E:CLS
60 G = 6.6732E-11: M = 1.989E30
70 P = 6.283185*((A*1.496E11)^(3/2))/SQR(G*M)
80 D = P/480: VX = 0
90 VY = SQR((1+E)*G*M/((1-E)*A*1.496E11))
100 X = (1-E)*A*1.496E11: Y = 0
110 X(0) = X/1.496E11:Y(0)=Y:T=1:K=1/2
120 PRINT"TIME<PERIOD/12>   ANG MOM          AREA<SQ M>":PRINT
130 FOR J = 1 TO 40:R= (X*X+Y*Y)^(3/2)
140 AX = -G*M*X/R: AY = -G*M*Y/R
150 VX = VX + AX*D*K: VY = VY + AY*D*K
160 X = X + VX*D: Y = Y + VY*D
170 Q(T) = Q(T) + (X*VY-Y*VX)*D/2
180 K = 1: NEXT J: B(T) = Q(T):L(T) = (X*VY-Y*VX)/1E15
190 X(T) = X/1.496E11: Y(T) = Y/1.496E11
200 PRINTTAB(3) T, L(T),B(T)
210 T = T+1:  IF T = 13 THEN 230
220 GOTO 130
230 PRINT:PRINT"ONE ORBIT COMPLETE. PERIOD = ";P/3.156E7;" YR":PRINT
240 PRINT"SEMI-MAJOR AXIS = ";A:PRINT:PRINT"ECCENTRICITY = ";E:PRINT
250 FOR J = 1 TO 1000:NEXT J
260 PRINT"IF YOU WANT TO SEE THE ORBITAL POSITIONS":PRINT
270 PRINT"ENTER 'GO'. TO QUIT, ENTER 'BYE'":PRINT
280 INPUT A$: IF A$= "GO" THEN 300
290 END
300 CLS:PRINT" TIME<P/12>       X<AU>           Y<AU>":PRINT
310 FOR J=0 TO 12:PRINT TAB(3) J,X(J),Y(J):NEXT J:END
```

```
10  REM  ** APPLE **
20  REM  ** KEPLER **
30  DIM B(15),Q(15): INPUT "INPUT SEMI-MAJOR AXIS IN AU ";A
40  PRINT : INPUT "INPUT ECCENTRICITY ";E: HOME
50 G = 6.6732E - 11:M = 1.989E30
60 P = 6.283185 * ((A * 1.496E11) ^ (3 / 2)) /  SQR (G * M)
70 D = P / 480:VX = 0
80 VY =  SQR ((1 + E) * G * M / ((1 - E) * A * 1.496E11))
90 X = (1 - E) * A * 1.496E11:Y = 0
100 T = 0:K = 1 / 2
110  PRINT "T<P/12>  X<AU>  Y<AU>   L    A<SQ M>": PRINT
120  GOTO 270
130  FOR J = 1 TO 40:R = (X * X + Y * Y) ^ (3 / 2)
140 AX =  - G * M * X / R:AY =  - G * M * Y / R
150 VX = VX + AX * D * K:VY = VY + AY * D * K
160 X = X + VX * D:Y = Y + VY * D
170 Q(T) = Q(T) + (X * VY - Y * VX) * D / 2
180 K = 1: NEXT J:PO =  INT ( LOG (Q(T)) / 2.30258509)
190 B(T) = ( INT (Q(T) / 10 ^ (PO - 3))) / 1000
200 L =  INT (X / 1.496E9) / 100:N =  INT (Y / 1.496E9) / 100
210 S =  INT ((X * VY - Y * VX) / 1E13) / 100
220  PRINT  TAB( 3)T;"      ";L;"     ";N;"   ";S;"  ";B(T);"E+";PO
230 T = T + 1:Q(T) = 0: IF T = 13 THEN 250
240  GOTO 130
250  PRINT : PRINT "ORBIT COMPLETE. PERIOD = ";P / 3.156E7;" YR": PRINT
260  PRINT "SEMI-MAJOR AXIS = ";A;" AU": PRINT "ECCENTRICITY = ";E: END
270 L =  INT (X / 1.496E9) / 100:N =  INT (Y / 1.496E9) / 100
280 S =  INT ((X * VY - Y * VX) / 1E13) / 100
290  PRINT  TAB( 3)T;"      ";L;"      ";N;"    ";S
300 T = T + 1: GOTO 130
```

The figure shows a spring-mass oscillator. In the absence of any frictional forces the displacement of the mass is described by Eq 14.4,

$$m(d^2x/dt^2) = -kx \qquad (14.4)$$

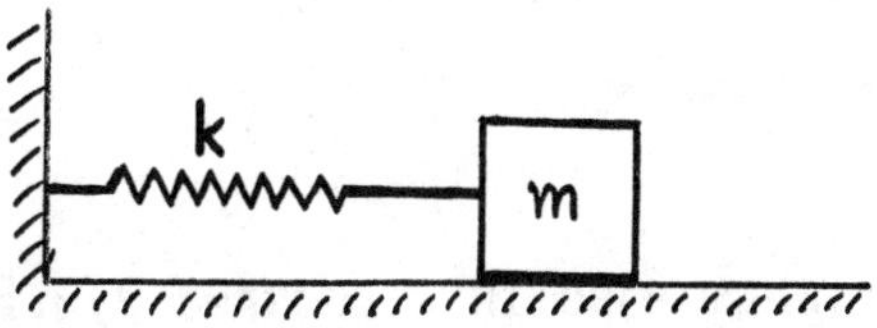

If the mass is subject to a frictional force ($-\gamma dx/dt$) in addition to the spring force then Eq 14.40 describes the motion,

$$m(d^2x/dt^2) = -\gamma dx/dt - kx \qquad (14.40)$$

The parameter γ measures the strength of the frictional damping force. You can use the program DAMP to study undamped oscillations simply by setting $\gamma = 0$. You can also use the program to study the transition from underdamped motion to critically damped motion, and then to overdamped motion.

The program starts the oscillator with an initial amplitude $x(t=0) = 10$, and an initial velocity $(dx/dt)_{t=0} = 0$. This corresponds to starting the motion by stretching the spring by 10 units and releasing it from rest.

With $\gamma = 0$ the oscillations are undamped. The resulting simple harmonic motion has a period

$$T = 2\pi/\omega = 2\pi[m/k]^{\frac{1}{2}}$$

The program integrates the equation of motion (Eq 14.4 or 14.40) and displays the position at intervals of 0.10 times the undamped period, $T = 2\pi[m/k]^{\frac{1}{2}}$.

You can use the program to solve Problems 14.3, 14.6, and 14.8 (or to check your analytic solutions) by choosing different values of the mass (Problem 14.3) or spring constant (14.6, 14.8) and running the program repeatedly until you obtain the desired period. Problems 14.40 and 14.41 may be studied by running the program with the appropriate values of m, k, and γ.

If you wish to study the transition from underdamped, to critically damped, to overdamped motion you can input the following values for m and k:

$$m = 1 \text{ kg}; \quad k = 1 \text{ N/m}$$

Critical damping then corresponds to $\gamma = 2$ kg/s. For $\gamma < 2$ kg/s you should find that the motion is oscillatory. For $\gamma > 2$ kg/s you should find that the motion "decays" without oscillation.

```
10 ** RADIO SHACK AND IBM PC **
20 REM ** DAMP **
30 CLS
40 INPUT"INPUT MASS IN KILOGRAMS ";M:PRINT
50 INPUT"INPUT SPRING CONSTANT IN NEWTONS/METER  ";K:PRINT
60 INPUT"INPUT GAMMA IN KILOGRAMS/SECOND ";G:PRINT
70 T = 6.28318*SQR(M/K)
80 CLS
90 PRINT"IF YOU WANT TO STOP THE PROGRAM":PRINT
100 PRINT"PRESS 'BREAK'. TO RESUME TYPE ":PRINT
110 PRINT"'CONTINUE' AND PRESS 'ENTER'":PRINT
120 FOR J = 1 TO 4000: NEXT J
130 PRINT"UNDAMPED PERIOD = ";T;"S":PRINT
140 D = T/1000
150 PRINT" TIME <S>        DISPLACEMENT ":PRINT
160 X0 = 10
170 PRINT Y,X0
180 X1 = -D*X0/2
190 Q = 1/2
200 FOR J = 1 TO 100
210 X2 = -K*X0/M - G*X1/M
220 X1 = X1 + X2*D*Q
230 X0 = X0 + X1*D
240 Y = Y + D
250 Q = 1
260 NEXT J
270 PRINT Y,X0
280 GOTO 200
```

```
10  REM  ** APPLE **
20  REM  ** DAMP **
30  HOME
40  INPUT "INPUT MASS IN KG ";M: PRINT : PRINT
50  INPUT "INPUT SPRING CONSTANT IN N/M ";K: PRINT : PRINT
60  INPUT "INPUT GAMMA IN KG/S ";G: PRINT
70 T = 6.28318 *  SQR (M / K)
80  HOME
90  PRINT "IF YOU WANT TO STOP THE PROGRAM": PRINT
100  PRINT "PRESS 'CONTROL', THEN 'C'": PRINT
110  PRINT "TO RESUME, ENTER 'CONT'": PRINT
120  FOR J = 1 TO 4000: NEXT J
130  PRINT "UNDAMPED PERIOD = ";T;" S": PRINT
140 D = T / 1000
150  PRINT " TIME<S>       DISPLACEMENT": PRINT
160 X0 = 10: PRINT Y,X0
170 X1 =  - D * X0 / 2
180 Q = 1 / 2
190  FOR J = 1 TO 100
200 X2 =  - K * X0 / M - G * X1 / M
210 X1 = X1 + X2 * D * Q
220 X0 = X0 + X1 * D
230 Y = Y + D:Q = 1
240  NEXT J
250  PRINT Y,X0
260  GOTO 190
```

Figure 18.10 (p 356) and Figure 37.2 (p 704) show two different waveforms and indicate how they may be represented by a sum of sinusoidal waveforms. Such sums are called Fourier series, and they are especially useful in dealing with systems that are linear. Roughly speaking, a system is linear if its response is directly proportional to its input. For example, the driven spring-mass system of Section 14.6 is linear. If we double the amplitude of the force which drives the system then the amplitude of the oscillations doubles. The force is the input or the stimulus and the amplitude is the output or response to that stimulus. In terms of the idea of a stimulus and a response we can define a linear system as one for which the response to a sum of stimuli equals the sum of the responses to the individual stimuli.

It turns out that it is relatively easy to analyze the response of a linear system to a sinusoidal stimulus. For example, in Sec 14.6 we determined the displacement of the spring-mass system when it is driven by a sinusoidal force. What if the driving force is not sinusoidal? How would the spring-mass system respond if we applied a square wave type of driving force? Here is where a Fourier series can be useful. We first represent the square wave force by a sum of sinusoidal forces. We then determine the response of the system to each of the sinusoidal forces. The sum of these responses gives us the response of the system to the sum of sinusoidal forces - the square wave type of force.

The programs TRIANGLE and SQUARE allow you to compare the value of a sum of sinusoidal waveforms with the exact value of the waveform. You get to choose the number of terms in the Fourier series. You should find that the series becomes a better approximation as you increase the number of terms. The triangular waveform of Fig 18.10 is represented by the Fourier series

$$T(x) = (8/\pi^2)\{\sin x - (1/9)\sin 3x + (1/25)\sin 5x - (1/49)\sin 7x + \ldots$$

The exact form of the function which T(x) represents is

$$T(x) = \begin{cases} 2x/\pi & 0 \leq x \leq \pi/2 \\ 2-2x/\pi & \pi/2 \leq x \leq \pi \end{cases}$$

The Fourier series which represents the square wave of Figure 37.2 is

$$S(x) = (4/\pi)\{\sin x + (1/3)\sin 3x + (1/5)\sin 5x + (1/7)\sin 7x + \ldots$$

The exact form of the function which S(x)
represents is

$$S(x) = \begin{cases} +1 & 0 < x < \pi \\ -1 & \pi < x < 2\pi \end{cases}$$

The programs TRIANGLE and SQUARE allow you to select and evaluate these two series. You choose the number of terms to be summed. The series is evaluated and displayed at ten positions over the interval 0 to π (for the triangular waveform) or from 0 to 2π (for the square waveform).

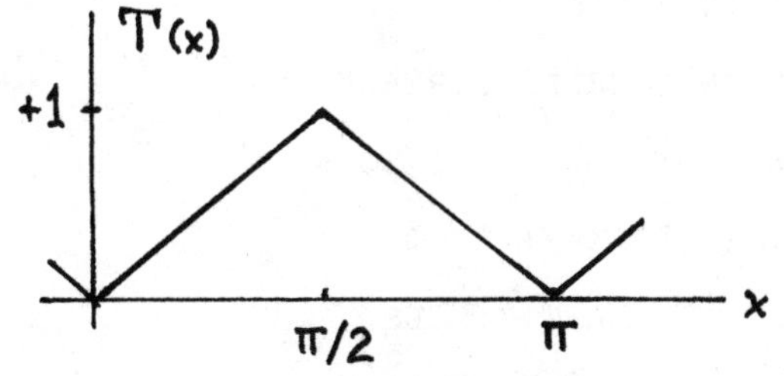

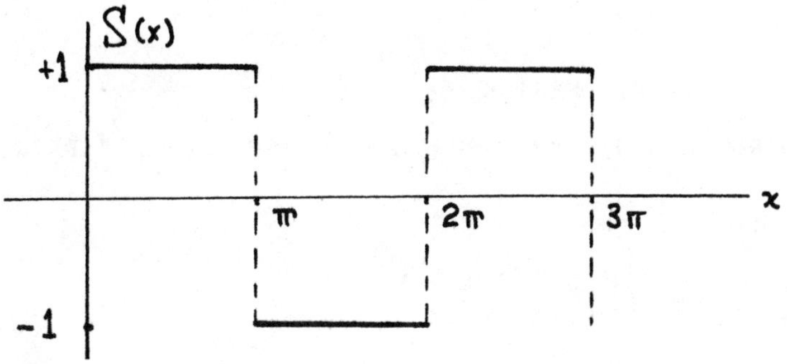

```
10 REM ** RADIO SHACK AND IBM PC **
20 REM ** SQUARE **
30 H = 2*3.14159/10: DIM F(101):CLS
40 INPUT"INPUT NUMBER OF TERMS IN FOURIER SERIES";N:PRINT
50 FOR K = 1 TO 10: X = K*H: S = 0: FOR J = 1 TO N
60 S = S + (4/3.14159)*SIN((2*J-1)*X)/(2*J-1)
70 NEXT J
80 F(K) = S: NEXT K: CLS
90 PRINT"NUMBER OF TERMS IN SERIES = ";N:PRINT:PRINT
100 PRINT"      X          F(X)":PRINT"   ====================="
110 FOR K = 1 TO 10
120 PRINT"  ";K*H;"   ";F(K):NEXT K
130 PRINT:PRINT"TO RUN AGAIN, TYPE 'RUN' AND PRESS 'ENTER'":PRINT
140 END
```

```
10 REM ** RADIO SHACK AND IBM PC **
20 REM ** TRIANGLE **
30 A = 1:H = 3.14159/10: DIM F(101):CLS
40 INPUT"INPUT NUMBER OF TERMS IN FOURIER SERIES";N:PRINT
50 FOR K = 1 TO 10: X = K*H: S = 0: FOR J = 1 TO N
60 S = S + A*(8/9.86959)*SIN((2*J-1)*X)/(2*J-1)^2
70 A = -A: NEXT J: F(K) = S: A = 1: NEXT K: CLS
80 PRINT"NUMBER OF TERMS IN SERIES = ";N:PRINT:PRINT
90 PRINT"      X          F(X)":PRINT"   ====================="
100 FOR K = 1 TO 10
110 PRINT"  ";K*H;"   ";F(K): NEXT K
120 PRINT:PRINT"TO RUN AGAIN, TYPE 'RUN' AND PRESS 'ENTER'":PRINT
130 END
```

```
10  REM  ** APPLE **
20  REM  ** SQUARE **
30 H = 2 * 3.14159 / 10: DIM F(101): HOME
40  INPUT "INPUT # TERMS IN FOURIER SERIES ";N: PRINT
50  FOR K = 1 TO 10:X = K * H:S = 0: FOR J = 1 TO N
60 S = S + (4 / 3.14159) *  SIN ((2 * J - 1) * X) / (2 * J - 1)
70  NEXT J
80 F(K) = S: NEXT K: HOME
90  PRINT "NUMBER OF TERMS IN SERIES = ";N: PRINT : PRINT
100  PRINT "    X               F(X)"
110  PRINT "==========================="
120  FOR K = 1 TO 10
130  PRINT K * H,F(K): NEXT K
140  PRINT : PRINT "TO RUN AGAIN ENTER 'RUN'": PRINT
150  END
```

```
10  REM  ** APPLE **
20  REM  ** TRIANGLE **
30 A = 1:H = 3.14159 / 10: DIM F(101): HOME
40  INPUT "INPUT # TERMS IN FOURIER SERIES ";N: PRINT
50  FOR K = 1 TO 10:X = K * H:S = 0: FOR J = 1 TO N
60 S = S + A * (8 / 9.86959) *  SIN ((2 * J - 1) * X) / (2 * J - 1) ^ 2
70 A =  - A: NEXT J:F(K) = S:A = 1: NEXT K: HOME
80  PRINT "NUMBER OF TERMS IN SERIES = ";N: PRINT : PRINT
90  PRINT "    X               F(X)"
100  PRINT "==========================="
110  FOR K = 1 TO 10
120  PRINT K * H,F(K): NEXT K
130  PRINT : PRINT "TO RUN AGAIN ENTER 'RUN'": PRINT
140  END
```

The speed of deep water waves is given by Eq 18.21.

$$v = [g\lambda/2\pi + 2\pi S/\rho\lambda]^{\frac{1}{2}} \qquad (18.21)$$

The acceleration of gravity g and the surface tension S characterize the two restoring forces that act on water that is displaced in wave motion. For very short wavelengths the waves are called ripples and surface tension is dominant. For long wavelengths the gravitational force is dominant. Equation 18.21 shows that v becomes large at both very short and very long wavelengths. As the figure suggests the wavespeed has a minimum at some intermediate wavelength.

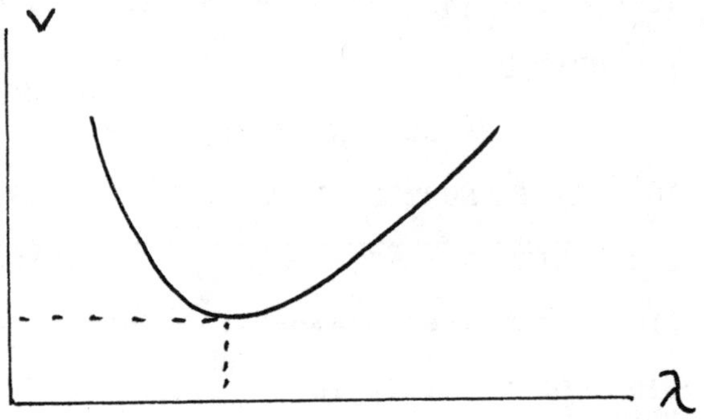

The program WAVES locates the minimum wavespeed by comparing the wave speed at one wavelength with the speed at slightly longer and slightly shorter wavelengths. The program identifies the minimum wavespeed as the one for which the speed at both of these adjacent wavelengths is greater than the test wavelength. The program then evaluates and displays both the minimum wavespeed and the wavelength for which the minimum occurs. As a final step the program tabulates and displays the wavespeed -vs- wavelength relation for speeds near the minimum.

The search technique is useful in many situations where you must locate a minimum or a maximum or a saddle point. The technique is used again in two programs dealing with electrostatics (See entries 11 and 12 in the directory on p 98 of this manual).

```
10 REM ** RADIO SHACK AND IBM PC **
20 REM ** WAVES **
30 S = 0.07: G = 9.8: A = 6.28318: R = 1E3: H = 0.001
40 DEF FNV(J) = G*J*H/A + (A*S/R)/(J*H): CLS
50 PRINT:PRINT"WATER WAVES EXHIBIT A MINIMUM SPEED FOR":PRINT
60 PRINT"SHORT WAVELENGTH WAVES CALLED RIPPLES.":PRINT
70 PRINT"THIS PROGRAM LOCATES THE MINIMUM WAVESPEED":PRINT
80 PRINT"AND TABULATES WAVESPEEDS NEAR THE MINIMUM":PRINT
90 PRINT"AS A FUNCTION OF WAVELENGTH.":PRINT
100 FOR J=1 TO 6000:NEXT J:FOR J = 2 TO 1000
110 GOSUB 240
120 IF FNV(J)<FNV(J-1) AND FNV(J)<FNV(J+1) THEN 140
130 NEXT J: END
140 CLS: K = J: PRINT
150 M = INT(10000*((4*G*S/R)^(1/4) + 0.00005))/10000
160 PRINT"MINIMUM SPEED = ";100*M;" CM/S AT WAVELENGTH = ";100*K*H;" CM":PRINT
170 FOR J = 1 TO 1500: NEXT J
180 PRINT"PRINTED BELOW ARE WAVESPEEDS NEAR THE MINIMUM":PRINT
190 FOR J = 1 TO 2000:NEXT J
200 PRINT"WAVELENGTH <CM>  WAVESPEED <CM/S>":PRINT
210 FOR J = K-4 TO K+4
220 PRINT TAB(5) J*H*100,TAB(19) INT(10000*(SQR(FNV(J))+0.00005))/100
230 NEXT J:PRINT:END
240 CLS:IF K = 1 THEN 280
250 IF K = 1 THEN 280
260 PRINT @ 350, "SEARCHING"
270 FOR M=1 TO 70:NEXT M: K = 1: RETURN
280 PRINT @ 350, ""
290 FOR M=0 TO 70:NEXT M: K = 0: RETURN
```

```
10  REM  ** APPLE **
20  REM  ** WAVES **
30 S = 0.07:G = 9.8:A = 6.28318:R = 1E3:H = 0.001
40  DEF  FN V(J) = G * J * H / A + (A * S / R) / (J * H): HOME
50  PRINT : PRINT "WATER WAVES EXHIBIT A MINIMUM SPEED FOR": PRINT
60  PRINT "SHORT WAVELENGTH WAVES CALLED RIPPLES.": PRINT
70  PRINT "THIS PROGRAM FINDS THE MINIMUM SPEED": PRINT
80  PRINT "AND TABULATES WAVESPEEDS NEAR THE": PRINT
90  PRINT "MINIMUM AS A FUNCTION OF WAVELENGTH.": PRINT
100  FOR J = 1 TO 6000: NEXT J: FOR J = 2 TO 1000
110  GOSUB 250
120  IF  FN V(J) <  FN V(J - 1) AND  FN V(J) <  FN V(J + 1) THEN 140
130  NEXT J: END
140  HOME :K = J: PRINT
150 M =  INT (1E4 * ((4 * G * S / R) ^ (1 / 4) + .00005)) / 1E4
160  PRINT "V(MIN) = ";100 * M;" CM/S,  L = ";100 * K * H;" CM": PRINT
170  FOR J = 1 TO 1500: NEXT J
180  PRINT "WAVESPEEDS NEAR THE MINIMUM": PRINT
190  FOR J = 1 TO 2000: NEXT J
200  PRINT "WAVELENGTH<CM>   WAVESPEED<CM/S>": PRINT
210  FOR J = K - 4 TO K + 4
220 SP =  SQR ( FN V(J)) + 0.00005:SP =  INT (1E4 * SP) / 100
230  PRINT  TAB( 5)100 * J * H, TAB( 19)SP
240  NEXT J: PRINT : END
250  HOME : IF K = 1 THEN 290
260  IF K = 1 THEN 290
270  HOME : PRINT "SEARCHING"
280  FOR M = 1 TO 100: NEXT M:K = 1: RETURN
290  HOME : PRINT
300  FOR M = 0 TO 100: NEXT M:K = 0: RETURN
```

The Maxwellian distribution function given by Eq 25.40 describes a gas of N identical molecules of mass m at a temperature T.

$$f(v) = 4\pi N(m/2\pi kT)^{3/2}v^2\exp(-\tfrac{1}{2}mv^2/kT) \qquad (25.40)$$

Noting that $k = R/N_A$ lets us convert Eq 25.40 to

$$f(v) = 4\pi N(M/2\pi RT)^{3/2}v^2\exp(-\tfrac{1}{2}Mv^2/RT)$$

where $M = mN_A$ is the molecular 'weight' (mass).

The program MAX first evaluates f(v) over a range of speeds which straddle the most probable speed,

$$v_{\text{most probable}} = [2kT/m]^{\frac{1}{2}}$$

You input the number of molecules (N), the molecular weight (M, expressed in grams per mole), and the kelvin temperature (T). The program then evaluates f(v) over a range of speeds which you specify. The program also enables you to evaluate the integral

$$\int_{v_1}^{v_2} f(v)dv \quad = \text{number of molecules in range } v_1 \text{ to } v_2$$

The value of the integral is displayed along with the fraction of all molecules with speeds in the range v_1 to v_2. If you choose a value for v_2 that is more than about 10 times the most probable speed the integral may not be accurate because the Simpson rule evaluation divides the range v_2-v_1 into 200 segments of equal length. If v_2-v_1 is too large the numerical integration will lack precision.

As a final option you can evaluate f(v) for speeds of your choice. This part of the program is an endless loop.

The program MAX should be helpful with Problems 25.22, 25.23, 25.25, 25.26, and 25.27.

```
10 REM ** RADIO SHACK AND IBM PC **
20 REM ** MAX **
30 DEF FNF(V) = A*V*V*EXP(-B*V*V)
40 CLS:INPUT"INPUT NUMBER OF MOLECULES ";N:PRINT
50 INPUT"INPUT MOLECULAR WEIGHT IN GRAMS ";M:PRINT
60 INPUT"INPUT KELVIN TEMPERATURE ";T:PRINT
70 A = 12.56637*N*((M/(5.2241E4*T))^(3/2))
80 B = M/(1.66288E4*T): P = SQR(1/B)
90 CLS:PRINT"THE PROGRAM WILL NOW EVALUATE AND ":PRINT
100 PRINT"DISPLAY F(V) FOR A RANGE OF SPEEDS THAT ":PRINT
110 PRINT"STRADDLE THE MOST PROBABLE SPEED":PRINT
120 FOR J = 1 TO 3000: NEXT J
130 FOR J = 1 TO 6: V = (2*J-1)*P/11
140 PRINT"F(";INT(V+1/2);") = ";FNF(V):NEXT J
150 FOR J = 1 TO 7: V = V+P/4
160 PRINT"F(";INT(V+1/2);") = ";FNF(V): NEXT J
170 PRINT:PRINT"WHEN YOU ARE READY TO CONTINUE PRESS ANY KEY":PRINT
180 A$ = INKEY$
190 IF A$ = "" THEN 180
200 CLS:PRINT:PRINT"THIS SEGMENT EVALUATES F(V) FOR 13 VALUES OF V.":PRINT
210 PRINT"YOU WILL BE PROMPTED TO INPUT THE LOWER AND ":PRINT
220 PRINT"UPPER LIMITS OF THE RANGE OF SPEEDS.":PRINT
230 FOR J =1 TO 5000: NEXT J:PRINT
240 PRINT"THE MOST PROBABLE SPEED WILL BE DISPLAYED AS A ":PRINT
250 PRINT"GUIDE. ALL SPEEDS ARE IN METERS/SECOND.":PRINT
260 FOR J = 1 TO 5000:NEXT J
270 PRINT:PRINT"MOST PROBABLE SPEED = ";INT(P+1/2);"M/S":PRINT
280 FOR J =1 TO 3000:NEXT J
```

CONTINUES ON NEXT PAGE

```
290 INPUT"INPUT LOWER LIMIT SPEED ";V1:PRINT
300 INPUT"INPUT UPPER LIMIT SPEED ";V2:PRINT:CLS
310 D = (V2-V1)/12: FOR K = 0 TO 12: V = V1 + D*K
320 PRINT"F(";INT(V+1/2);") = ";FNF(V):NEXT K
330 PRINT:PRINT"IF YOU WANT TO INTEGRATE F(V), TYPE AND":PRINT
340 PRINT"ENTER 'SUM'. TO EVALUATE F(V) FOR SPECIFIC":PRINT
350 PRINT"SPEEDS OF YOUR CHOICE, TYPE AND ENTER 'EVAL'":PRINT
360 INPUT B$:IF B$= "EVAL" THEN 530
370 PRINT:PRINT"YOU WILL BE PROMPTED TO ENTER THE LOWER":PRINT
380 PRINT"LIMIT SPEED AND THE UPPER LIMIT SPEED.":PRINT
390 FOR J = 1 TO 2000:NEXT J
400 INPUT"INPUT LOWER LIMIT SPEED ";V1:PRINT
410 INPUT"INPUT UPPER LIMIT SPEED ";V2:PRINT: H = (V2-V1)/400
420 S = H*(FNF(V1) + 4*FNF(V1+H) + FNF(V2))/3
430 FOR J = 2 TO 398 STEP 2: V = V1 + J*H
440 S = S + 2*H*(FNF(V) + 2*FNF(V+H))/3: NEXT J:CLS
450 PRINT:PRINT"INTEGRAL OF F(V) FROM V1 = ";V1:PRINT
460 PRINT"TO V2 = ";V2;" EQUALS ";INT(S+1/2):PRINT
470 PRINT"TOTAL NUMBER OF MOLECULES IS ";N:PRINT
480 PRINT"FRACTION IN RANGE V1 TO V2 IS ";INT(S+1/2)/N:PRINT:PRINT
490 PRINT:PRINT"TO EVALUATE F(V) FOR SPECIFIC SPEEDS OF YOUR":PRINT
500 PRINT"CHOICE, TYPE AND ENTER 'EVAL'. TO QUIT":PRINT
510 PRINT"TYPE AND ENTER 'Q'":PRINT
520 INPUT L$:IF L$= "Q" THEN 600
530 CLS:PRINT:PRINT"THIS SEGMENT IS AN ENDLESS LOOP. YOU WILL":PRINT
540 PRINT"BE PROMPTED TO ENTER A SPEED. THE PROGRAM":PRINT
550 PRINT"EVALUATES AND DISPLAYS F(V) AND THEN PROMPTS":PRINT
560 PRINT"YOU FOR ANOTHER SPEED. HAVE FUN!":PRINT
570 INPUT"INPUT SPEED ";V
580 PRINT:PRINT"F(";INT(V+1/2);") = ";FNF(V):PRINT
590 GOTO 570
600 END
```

```
10  REM  ** APPLE **
20  REM  ** MAX **
30  DEF  FN F(V) = A * V * V *  EXP ( - B * V * V): HOME
40  INPUT "INPUT NUMBER OF MOLECULES ";N: PRINT
50  INPUT "INPUT MOLECULAR WEIGHT IN GRAMS ";M: PRINT
60  INPUT "INPUT KELVIN TEMPERATURE ";T: PRINT
70 A = 12.56637 * N * ((M / (5.2241E4 * T)) ^ (3 / 2))
80 B = M / (1.66288E4 * T):P =  SQR (1 / B)
90  HOME : PRINT "THE PROGRAM WILL NOW EVALUATE AND": PRINT
100  PRINT "DISPLAY F(V) FOR A RANGE OF SPEEDS THAT": PRINT
110  PRINT "STRADDLE THE MOST PROBABLE SPEED": PRINT
120  FOR J = 1 TO 3000: NEXT J
130  FOR J = 1 TO 6:V = (2 * J - 1) * P / 11
140  PRINT "F("; INT (V + 1 / 2);") = "; FN F(V): NEXT J
150  FOR J = 1 TO 7:V = V + P / 4
160  PRINT "F("; INT (V + 1 / 2);") = "; FN F(V): NEXT J
170  PRINT : PRINT "TO CONTINUE, ENTER 'GO'": PRINT
180  INPUT G$: IF G$ = "GO" THEN 200
190  GOTO 180
200  HOME : PRINT "THIS SEGMENT EVALUATES F(V) FOR 13": PRINT
210  PRINT "VALUES OF V OVER A RANGE THAT YOU CHOOSE": PRINT
220  PRINT "YOU WILL BE PROMPTED TO INPUT THE LOWER": PRINT
230  PRINT "AND UPPER LIMITS OF THE RANGE OF SPEEDS": PRINT
240  FOR J = 1 TO 5000: NEXT J: PRINT
250  PRINT "THE MOST PROBABLE SPEED IS DISPLAYED": PRINT
260  PRINT "AS A GUIDE. ALL SPEEDS ARE IN M/S": PRINT
270  FOR J = 1 TO 5000: NEXT J
280  PRINT : PRINT "MOST PROBABLE SPEED = "; INT (P + 1 / 2);" M/S": PRINT
```

CONTINUES ON NEXT PAGE

```
290  FOR J = 1 TO 3000: NEXT J
300  INPUT "INPUT LOWER LIMIT SPEED ";V1: PRINT
310  INPUT "INPUT UPPER LIMIT SPEED ";V2: PRINT : HOME
320 D = (V2 - V1) / 12: FOR K = 0 TO 12:V = V1 + D * K
330  PRINT "F("; INT (V + 1 / 2);") = "; FN F(V): NEXT K
340  PRINT : PRINT "IF YOU WANT TO INTEGRATE F(V),": PRINT
350  PRINT "TYPE AND ENTER 'SUM'. TO EVALUATE": PRINT
360  PRINT "F(V) FOR SPECIFIC SPEEDS YOU CHOOSE": PRINT
370  PRINT "TYPE AND ENTER 'EVAL'": PRINT
380  INPUT B$: IF B$ = "EVAL" THEN 600
390  PRINT : PRINT "YOU WILL BE PROMPTED TO ENTER THE": PRINT
400  PRINT "LOWER AND UPPER LIMIT SPEEDS": PRINT
410  FOR J = 1 TO 2000: NEXT J: INPUT "INPUT LOWER LIMIT SPEED ";V1: PRINT
420  INPUT "INPUT UPPER LIMIT SPEED ";V2: PRINT :H = (V2 - V1) / 400
430 S = H * ( FN F(V1) + 4 *  FN F(V1 + H) +  FN F(V2)) / 3
440  FOR J = 2 TO 398 STEP 2:V = V1 + J * H
450 S = S + 2 * H * ( FN F(V) + 2 *  FN F(V + H)) / 3: NEXT J
460  HOME : PRINT : PRINT "INTEGRAL OF F(V) FROM V1 = ";V1: PRINT
470  PRINT "TO V2 = ";V2" EQUALS "; INT (S + 1 / 2): PRINT
480  PRINT "TOTAL NUMBER OF MOLECULES IS ";N: PRINT
490  PRINT "FRACTION IN RANGE V1 TO V2 IS "; INT (S + 1 / 2) / N: PRINT : PRINT
500  PRINT : PRINT "TO EVALUATE F(V) FOR SPEEDS OF": PRINT
510  PRINT "YOUR CHOICE, TYPE AND ENTER 'EVAL'": PRINT
520  PRINT "TO QUIT, TYPE AND ENTER 'Q'": PRINT
530  INPUT L$: IF L$ = "Q" THEN 600
540  HOME : PRINT : PRINT "THIS SEGMENT IS AN ENDLESS LOOP.": PRINT
550  PRINT "YOU WILL BE PROMPTED TO ENTER A SPEED": PRINT
560  PRINT "THE PROGRAM EVALUATES AND DISPLAYS F(V)": PRINT
570  PRINT "HAVE FUN!": PRINT
580  INPUT "INPUT SPEED ";V
590  PRINT : PRINT "F(" INT (V + 1 / 2);") = "; FN F(V): PRINT : GOTO 580
600  END
```

Problems 26.36 and 27.11 deal with the arrangement of charges shown in the figure. Four identical charges are located on the corners of a square. How can we use the computer to locate the points where the electric field is zero? The components of the electric field in the x-y plane are related to the potential by

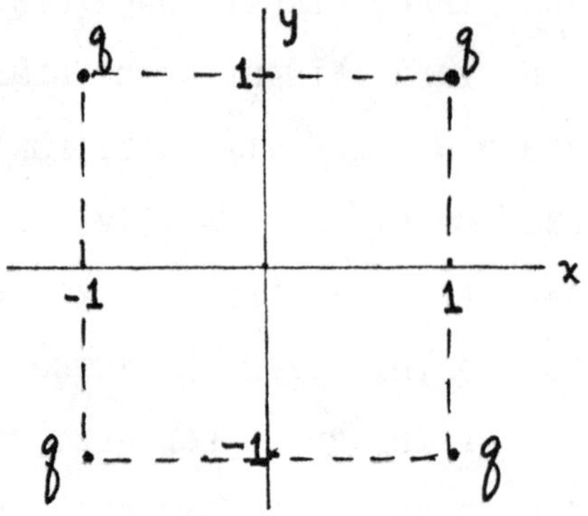

$$E_x = -\partial V/\partial x \ ; \quad E_y = -\partial V/\partial y$$

Positions where $E_x = 0$ are positions where the derivative $\partial V/\partial x = 0$. Positions where $E_y = 0$ are positions where $\partial V/\partial y = 0$. At points where $E_x = 0$ and $E_y = 0$ the potential $V(x,y)$ can be a minimum, a maximum, or a saddle point.

The program V4 systematically compares the potential at a test point with the potential at four nearby points - two along the x-axis and two along the y-axis. The program searches only over the quadrant $x \geqslant 0$, $y \geqslant 0$. Symmetry allows you to determine where E = 0 positions are located in the other quadrants.

The program V3 performs a similar search for E = 0 positions for the triangular array of charges shown at the right. The program searches over the right half of the triangle. Symmetry allows you to determine where E = 0 positions are located in the left half of the triangle.

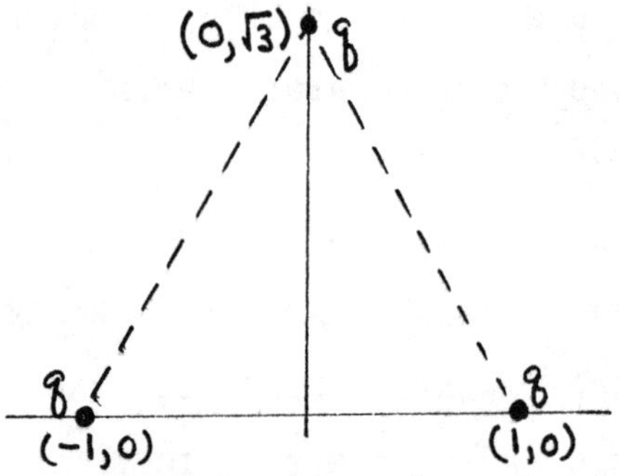

There are a total of five E = 0 positions for the square array of charges, and a total of four E = 0 positions for the triangular array. Make a guess now as to where the E = 0 positions are located. If you test your intuition by making a qualitative guess the computer solution will have a greater impact.

Both V4 and V3 require fairly long run times. The run time is proportional to the square of the number of segments you will be prompted to input. A small number of segments - say 8 - will give a run time of from two to four minutes.

There are no V3 or V4 listings for the Apple. There are Apple listings for the programs V3S and V4S described on page 143.

```
10 REM ** RADIO SHACK AND IBM PC **
20 REM ** V3 **
30 PRINT"THIS PROGRAM LOCATES E = 0 POSITIONS.":PRINT
40 PRINT"YOU WILL BE PROMPTED TO CHOOSE THE NUMBER":PRINT
50 PRINT"OF SEGMENTS. CHOOSE A VALUE > 7.":PRINT
60 PRINT"TRY 8 FOR STARTERS!":PRINT
70 INPUT"INPUT THE NUMBER OF STEPS";N:CLS:H=1/N
80 PRINT @ 220, "SEARCHING":PRINT:PRINT
90 DEF FNS(J,K) = 1/SQR((1-J*H)^2+(K*H)^2)
100 DEF FNT(J,K) = 1/SQR((1+J*H)^2+(K*H)^2)
110 DEF FNU(J,K) = 1/SQR((J*H)^2+(1.7320508-K*H)^2)
120 DEF FNV(J,K) = FNS(J,K)+FNT(J,K)+FNU(J,K)
130 FOR K = 0 TO N-2: FOR J = 0 TO N-2-K
140 IF FNV(J,K)>FNV(J-1,K) AND FNV(J,K)>FNV(J+1,K) THEN GOSUB 180
150 IF FNV(J,K)<FNV(J-1,K) AND FNV(J,K)<FNV(J+1,K) THEN GOSUB 210
160 NEXT J: NEXT K: IF K=N-1 THEN 170: GOTO 130
170 PRINT:PRINT"NUMBER OF STEPS = ";N;"   STEP SIZE = ";H:END
180 IF FNV(J,K)>FNV(J,K-1) AND FNV(J,K)>FNV(J,K+1) THEN GOSUB 240
190 IF FNV(J,K)<FNV(J,K-1) AND FNV(J,K)<FNV(J,K+1) THEN GOSUB 260
200 RETURN
210 IF FNV(J,K)<FNV(J,K-1) AND FNV(J,K)<FNV(J,K+1) THEN GOSUB 280
220 IF FNV(J,K)>FNV(J,K-1) AND FNV(J,K)>FNV(J,K+1) THEN GOSUB 300
230 RETURN
240 PRINT"E = O AT X = ";J*H;"  Y = ";K*H;"  V = ";FNV(J,K):PRINT
250 RETURN
260 PRINT"E = 0 AT X = ";J*H;" Y = ";K*H;"  V = ";FNV(J,K):PRINT
270 RETURN
280 PRINT"E = 0 AT X = ";J*H;"  Y = ";K*H;"  V = ";FNV(J,K):PRINT
290 RETURN
300 PRINT"E = 0 AT X = ";J*H;"  Y = ";K*H;"  V = ";FNV(J,K):PRINT
310 RETURN
```

```
10 REM ** RADIO SHACK AND IBM PC **
20 REM ** V4 **
30 CLS:PRINT"THIS PROGRAM LOCATES E = 0 POSITIONS.":PRINT
40 PRINT"YOU WILL BE PROMPTED TO CHOOSE THE NUMBER":PRINT
50 PRINT"OF SEGMENTS INTO WHICH THE SQUARE IS DIVIDED":PRINT
60 PRINT"CHOOSE A VALUE GREATER THAN 7. TRY 8 FOR STARTERS!":PRINT
70 INPUT"INPUT THE NUMBER OF STEPS";N:CLS: H=1/N
80 PRINT @ 220, "SEARCHING":PRINT:PRINT
90 DEF FNR(J,K) = 1/SQR((1-J*H)^2+(1-K*H)^2)
100 DEF FNS(J,K) = 1/SQR((1-J*H)^2+(1+K*H)^2)
110 DEF FNT(J,K) = 1/SQR((1+J*H)^2+(1-K*H)^2)
120 DEF FNU(J,K) = 1/SQR((1+J*H)^2+(1+K*H)^2)
130 DEF FNV(J,K) = FNR(J,K)+FNS(J,K)+FNT(J,K)+FNU(J,K)
140 FOR K = 0 TO N-2: FOR J=0 TO N-2
150 IF FNV(J,K)>FNV(J-1,K) AND FNV(J,K)>FNV(J+1,K) THEN GOSUB 190
160 IF FNV(J,K)<FNV(J-1,K) AND FNV(J,K)<FNV(J+1,K) THEN GOSUB 220
170 NEXT J:NEXT K:IF K = N-1 THEN 180: GOTO 140
180 PRINT:PRINT"NUMBER OF STEPS = ";N;"   STEP SIZE = ";H:END
190 IF FNV(J,K)>FNV(J,K-1) AND FNV(J,K)>FNV(J,K+1) THEN GOSUB 250
200 IF FNV(J,K)<FNV(J,K-1) AND FNV(J,K)<FNV(J,K+1) THEN GOSUB 270
210 RETURN
220 IF FNV(J,K)<FNV(J,K-1) AND FNV(J,K)<FNV(J,K+1) THEN GOSUB 290
230 IF FNV(J,K)>FNV(J,K-1) AND FNV(J,K)>FNV(J,K+1) THEN GOSUB 310
240 RETURN
250 PRINT"E = 0 AT X = ";J*H;"  Y = ";K*H;"  V = ";FNV(J,K):PRINT
260 RETURN
270 PRINT"E = 0 AT X = ";J*H;" Y = ";K*H;"  V = ";FNV(J,K):PRINT
280 RETURN
290 PRINT"E = 0 AT X = ";J*H;"  Y = ";K*H;"  V = ";FNV(J,K):PRINT
300 RETURN
310 PRINT"E = 0 AT X = ";J*H;"  Y = ";K*H;"  V = ";FNV(J,K):PRINT
320 RETURN
```

The programs V3S and V4S locate E = 0 positions for the triangle and square charge distributions analyzed by the programs V3 and V4. By exploiting symmetry V3S and V4S avoid the time-consuming searches used in V3 and V4.

For the distribution with four identical charges at the corners of a square, symmetry dictates that E = 0 positions be located on the x-axis, the y-axis, or on the diagonals x = y and x = -y. On the x-axis, $E_y = 0$, by symmetry. The program V4 searches for positions where $E_x = -\partial V/\partial x = 0$. The program locates positions where $\partial V/\partial x = 0$ by picking a test point and comparing the potential at that point with the potential at points on either side. If the potential at the test point is greater than the potential at the points on either side then the program concludes that the test point is a potential maximum where $\partial V/\partial x = 0$ and $E_x = 0$. The program also tests for a potential minimum, where $\partial V/\partial x = 0$ and $E_x = 0$. A similar search is made along the diagonal line x = y.

The symmetry of the triangular array of charges requires that E = 0 positions lie on the lines which bisect the angles at each vertex. The program V3S searches along the y-axis. As with V4S, the E = 0 positions are identified by comparing the potential at a test point with the potential at points immediately above and below.

Both V3S and V4S use one-dimensional searches, and require much shorter run times than the programs V3 and V4 which use two-dimensional searches. You might want to start by choosing a search which uses 20 segments. Then repeat with 100 segments to see how much the E = 0 locations are changed.

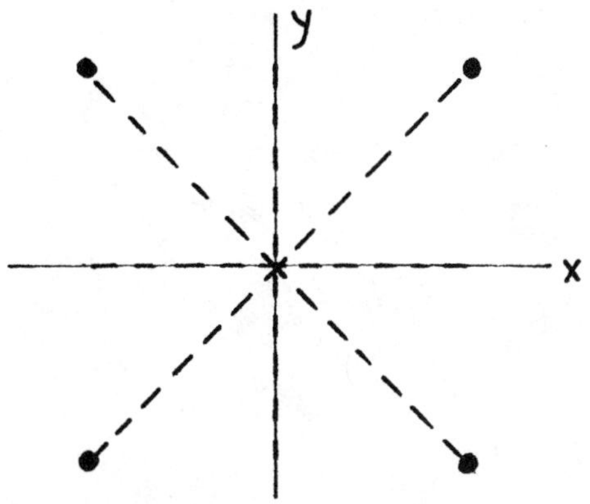

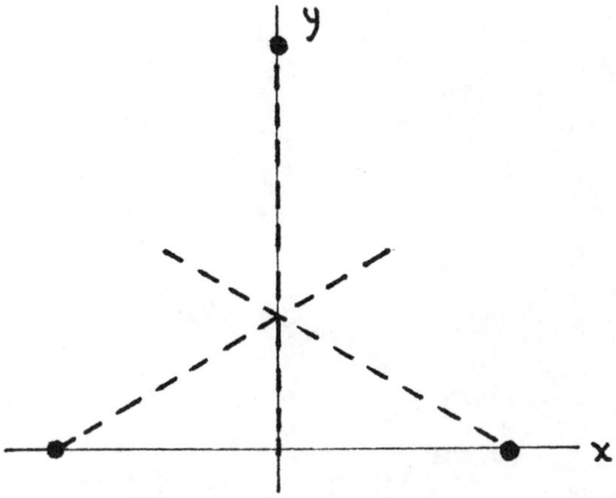

```
10 REM ** RADIO SHACK AND IBM PC **
20 REM ** V3S **
30 CLS:PRINT"THIS PROGRAM LOCATES E = 0 POSITIONS":PRINT
40 PRINT"ALONG THE Y-AXIS. YOU CHOOSE THE":PRINT
50 PRINT"NUMBER OF TEST POINTS ALONG THE":PRINT
60 PRINT"Y-AXIS. TRY 20 FOR STARTERS!":PRINT
70 INPUT"INPUT NUMBER OF TEST POINTS ";N:CLS:H = 1.7320508/N
80 DEF FNV(J) = 2/SQR(1 + (J*H)^2) + 1/(1.7320508-J*H)
90 PRINT:PRINT"SEARCHING ALONG Y-AXIS":PRINT
100 FOR J= 0 TO N-2
110 IF FNV(J)>FNV(J-1) AND FNV(J)>FNV(J+1) THEN 140
120 NEXT J
130 GOTO 150
140 PRINT"E = 0 AT Y = ";J*H:PRINT:NEXT J
150 FOR J=0 TO N-2
160 IF FNV(J)<FNV(J-1) AND FNV(J)<FNV(J+1) THEN 200
170 NEXT J
180 PRINT:PRINT"NUMBER OF STEPS = ";N;"   STEP SIZE = ";H:PRINT
190 PRINT"TO RUN AGAIN, ENTER 'RUN'.":PRINT:END
200 PRINT"E = 0 AT Y = ";J*H:PRINT:NEXT J
210 PRINT:PRINT"NUMBER OF STEPS = ";N;"   STEP SIZE = ";H:PRINT
220 PRINT"TO RUN AGAIN, ENTER 'RUN'.":PRINT:END
```

```
10 REM ** RADIO SHACK AND IBM PC **
20 REM ** V4S **
30 CLS:PRINT"THIS PROGRAM LOCATES E = 0":PRINT
40 PRINT"POSITIONS ALONG THE X-AXIS AND":PRINT
50 PRINT"ALONG THE DIAGONAL LINE X = Y.":PRINT:PRINT
60 INPUT"INPUT THE NUMBER OF TEST POINTS ";N:H=1/N
70 DEF FNV(J) = 2/SQR(1+(1+J*H)^2)+2/SQR(1+(1-J*H)^2)
80 CLS:PRINT"SEARCHING ALONG X-AXIS":PRINT:FOR J=1 TO 3000:NEXT J
90 FOR J= 0 TO N-2
100 IF FNV(J)>FNV(J-1) AND FNV(J)>FNV(J+1) THEN 130
110 NEXT J
120 GOTO 140
130 PRINT"E = 0 AT X = ";J*H:PRINT:NEXT J
140 FOR J = 0 TO N-2
150 IF FNV(J)< FNV(J-1) AND FNV(J)< FNV(J+1) THEN 190
160 NEXT J
170 IF Q = 1 THEN 250
180 GOTO 200
190 PRINT"E = 0 AT X = ";J*H:PRINT:NEXT J
200 PRINT"SEARCHING ALONG DIAGONAL X = Y":PRINT
210 DEF FNV(J) = 1.4142135/(1-(J*H)^2)+2/SQR((1+J*H)^2+(1-J*H)^2)
220 Q = 1
230 GOTO 90
240 PRINT:PRINT"NUMBER OF STEPS = ";N;"    STEP SIZE = ";H:PRINT
250 PRINT"TO RUN AGAIN, ENTER 'RUN'":PRINT:END
```

```
10  REM  ** APPLE **
20  REM  ** V3S **
30  HOME : PRINT "THIS PROGRAM LOCATES E = 0 POSITIONS": PRINT
40  PRINT "ALONG THE Y-AXIS. YOU CHOOSE THE": PRINT
50  PRINT "NUMBER OF TEST POINTS ALONG THE": PRINT
60  PRINT "Y-AXIS. TRY 20 FOR STARTERS! ": PRINT
70  INPUT "INPUT THE NUMBER OF TEST POINTS ";N: HOME :H = 1.7320508 / N
80  DEF  FN V(J) = 2 /  SQR (1 + (J * H)  ^ 2) + 1 / (1.7320508 - J * H)
90  FOR J = 0 TO N - 2
100  IF  FN V(J) >  FN V(J - 1) AND  FN V(J) >  FN V(J + 1) THEN 130
110  NEXT J
120  GOTO 140
130  PRINT "E = 0 AT Y = ";J * H: PRINT : NEXT J
140  FOR J = 0 TO N - 2
150  IF  FN V(J) <  FN V(J - 1) AND  FN V(J) <  FN V(J + 1) THEN 200
160  NEXT J
170 H =  INT (1E5 * H) / 1E5
180  PRINT : PRINT "NUMBER OF STEPS = ";N;"  STEP SIZE = ";H: PRINT
190  PRINT "TO RUN AGAIN, ENTER 'RUN'": PRINT : END
200  PRINT "E = 0 AT Y = ";J * H: PRINT : NEXT J
210 H =  INT (1E5 * H) / 1E5
220  PRINT : PRINT "NUMBER OF STEPS = ";N;"  STEP SIZE = ";H: PRINT
230  PRINT "TO RUN AGAIN, ENTER 'RUN' ": PRINT : END
```

```
10  REM  ** APPLE **
20  REM  ** V4S **
30  HOME : PRINT "THIS PROGRAM LOCATES E = 0": PRINT
40  PRINT "POSITIONS ALONG THE X-AXIS AND": PRINT
50  PRINT "ALONG THE DIAGONAL LINE X = Y.": PRINT : PRINT
60  INPUT "INPUT THE NUMBER OF TEST POINTS ";N:H = 1 / N
70  DEF  FN V(J) = 2 /  SQR (1 + (1 + J * H) ^ 2) + 2 /  SQR (1 + (1 - J * H) ^ 2)
80  HOME : PRINT "SEARCHING ALONG THE X-AXIS": PRINT : FOR J = 1 TO 3000: NEXT J
90  FOR J = 0 TO N - 2
100  IF  FN V(J) >  FN V(J - 1) AND  FN V(J) >  FN V(J + 1) THEN 130
110  NEXT J
120  GOTO 140
130  PRINT "E = 0 AT X = ";J * H: PRINT : NEXT J
140  FOR J = 0 TO N - 2
150  IF  FN V(J) <  FN V(J - 1) AND  FN V(J) <  FN V(J + 1) THEN 190
160  NEXT J
170  IF Q = 1 THEN 260
180  GOTO 200
190  PRINT "E = 0 AT X = ";J * H: PRINT : NEXT J
200  PRINT "SEARCHING ALONG DIAGONAL X = Y": PRINT
210  DEF  FN T(J) = 1.4142135 /  SQR (1 - (J * H) ^ 2)
220  DEF  FN U(J) = 2 /  SQR ((1 + J * H) ^ 2 + (1 - J * H) ^ 2)
230  DEF  FN V(J) =  FN T(J) +  FN U(J)
240 Q = 1
250  GOTO 90
260  PRINT : PRINT "NUMBER OF STEPS = ";N;"    STEP SIZE = ";H: PRINT
270  PRINT "TO RUN AGAIN, ENTER 'RUN'": PRINT : END
```

The infinite network of Problem 31.40 is shown in the figure. Let R denote the resistance of this network. We can evaluate R by developing an algorithm for calculating the resistance of finite networks like those shown. An algorithm is a method which uses a repetitive scheme of calculation. In this problem we can develop an algorithm that expresses the resistance for a network of N + 1 segments in terms of the resistance of a network of N segments. By setting N = 1,2,3,... in the algorithm we can evaluate the resistance for larger and larger networks. Intuitively we expect the resistance to approach a definite limit as N increases because each added segment adds relatively fewer 'paths' for the current to branch into. Indeed, as shown in Problem 31.40, the resistance of the infinite network must be less than 3r and greater than 2r.

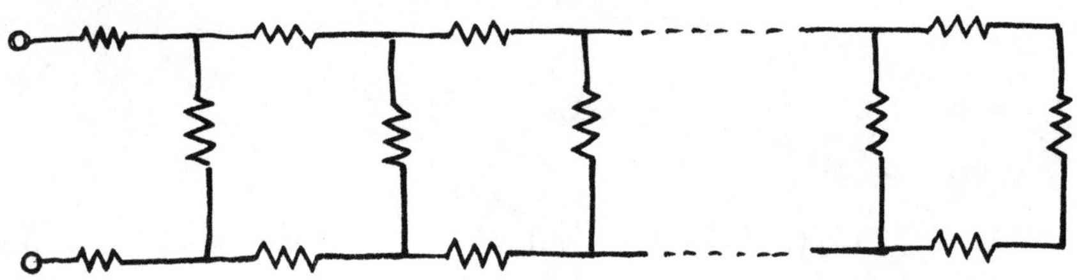

The infinite "ladder" network of Problem 31.40

If you have worked on Problem 31.40 know that the resistance of the infinite network is $(1 + \sqrt{3})r = 2.73205r$, where r is the resistance of the identical resistors making up the network. How large would you guess N must be before the finite network resistance reaches the infinite network resistance (to 6-figure accuracy)? Go ahead - make a guess! Then RUN the program. The "pictures" on the next page show you how to arrive at the algorithm

$$R(N+1) = 2r + rR(N)/(r + R(N))$$

In the computer program we have set r = 1, so the infinite network resistance is

$$R = 1 + \sqrt{3} = 2.73205$$

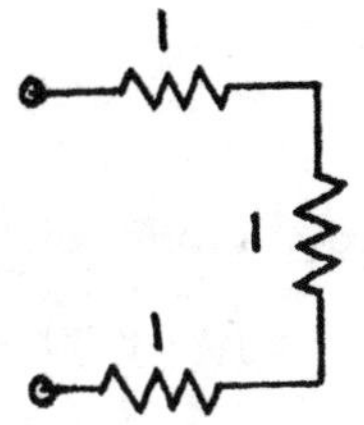

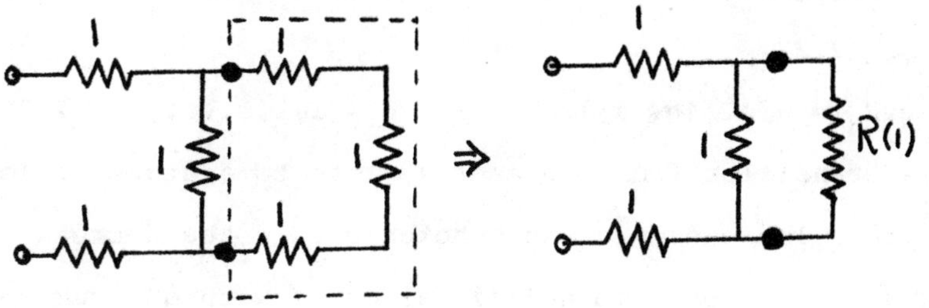

R(1) = 2 + 1 R(2) = 2 + 1•R(1)/(1 + R(1))

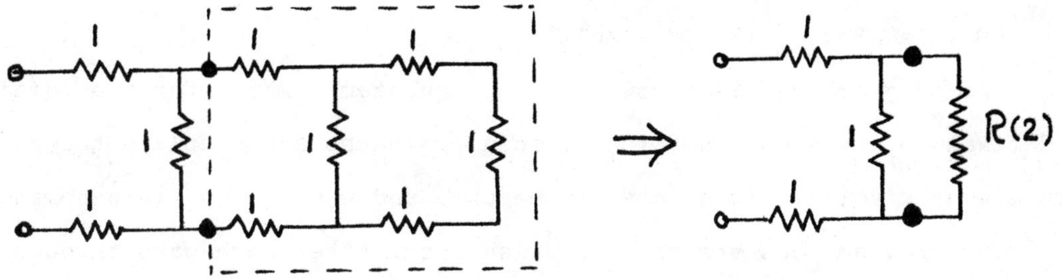

R(3) = 2 + 1•R(2)/(1 + R(2))

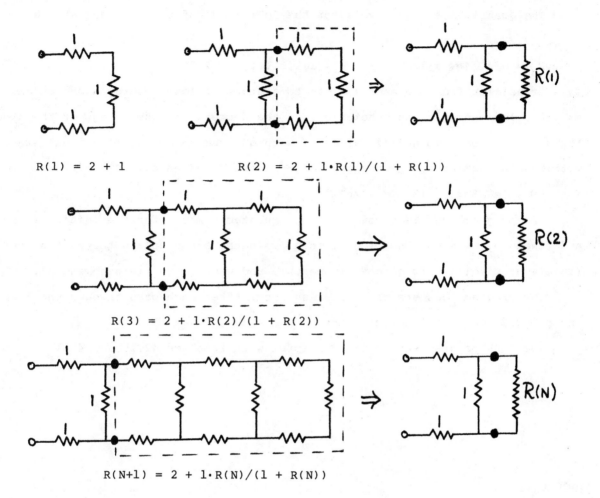

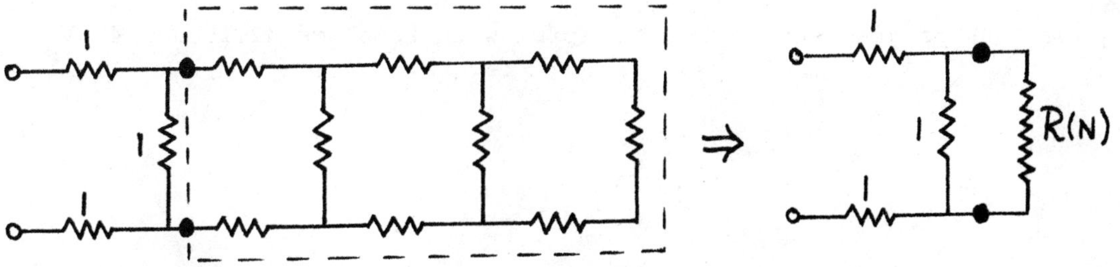

R(N+1) = 2 + 1•R(N)/(1 + R(N))

If you want to test this algorithm experimentally, take a group of equal resistances and construct the N = 1, 2, 3, etc. networks and measure the resistance of each with an ohmeter. Remember that even 'gold band' resistors are rated to within ± 5% of their nominal value.

The computer listing which follows - 1 line - will RUN on any of the three makes of computer.

```
10 INPUT"INPUT N";N:R=3:FOR J=2 TO N:R=2+R/(1+R):PRINT"N = ";J;" R = ";R:NEXT J:END
```

The quantized energy levels for the Bohr model of the hydrogen atom are given by Eq 42.22,

$$E_n = -[e^2/4\pi\epsilon_o a_o]/n^2 \qquad n = 1,2,3,\ldots \qquad (42.22)$$

In a transition from one energy state to a state of lower energy a photon is emitted. The energy of the photon equals the decrease in the energy of the atom. For a transition from an initial state with quantum number n_i to a final state with quantum number n_f the energy of the photon is given by

$$h\nu = [e^2/8\pi\epsilon_o a_o]\{1/n_f^2 - 1/n_i^2\}$$

The program BOHR lets you input the quantum numbers for the initial (n_i) and final (n_f) states. The program then evaluates and displays the photon energy (in electron volts), frequency (in hertz), and wavelength (in nanometers).

The program consists of an endless loop. After each pass through the loop the program recaps all calculations.

The BOHR program should prove helpful with Problems 42.10 and 42.11.

```
10 REM ** RADIO SHACK AND IBM PC **
20 REM ** BOHR **
30 DIM NI(30), NF(30), E(30), F(30), L(30): K = 1
40 P = 8.98755E9: A = 5.29177E-11
50 H = 6.62620E-34: C = 2.99793E8
60 J = 1.60219E-19
70 CLS
80 PRINT:INPUT"INPUT INITIAL STATE QUANTUM NUMBER ";NI:PRINT
90 INPUT"INPUT FINAL STATE QUANTUM NUMBER ";NF:PRINT
100 CLS
110 PRINT" NI  NF  ENERGY<EV>   FREQUENCY<HZ>   WAVELENGTH<NM> ":PRINT
120 DEF FNE(NI,NF) = (J*J*P/(2*A))*(1/(NF*NF) - 1/(NI*NI))
130 E = FNE(NI,NF)/J
140 F = FNE(NI,NF)/H
150 L = (C*1E9)/F
160 NI(K) = NI: NF(K) = NF: E(K) = E: F(K) = F: L(K) = L
170 FOR S = 1 TO K
180 PRINT NI(S);" ";NF(S);"  ";E(S);"    ";F(S);"     ";L(S)
190 NEXT S
200 K = K+1
210 GOTO 80
```

```
10  REM  ** APPLE **
20  REM  ** BOHR **
30  DIM NI(150),NF(150),E(150),PE(150),F(150),PF(150),L(150),PL(150):K = 1
40 P = 8.98755E9:A = 5.29177E - 11
50 H = 6.62620E - 34:C = 2.99793E8
60 J = 1.60219E - 19
70  HOME
80  PRINT : PRINT : INPUT "INPUT INITIAL STATE QUANTUM NUMBER ";NI: PRINT
90  INPUT "INPUT FINAL STATE QUANTUM NUMBER ";NF: PRINT
100  HOME
110  PRINT "NI NF   E<EV>     F<HZ>        LAMBDA<NM>": PRINT
120 EI = J * J * P / (2 * A * NI * NI):EF = EI * (NI / NF) ^ 2
130 E = (EF - EI) / J
140 F = E * J / H
150 L = (C * 1E9) / F
160 NI(K) = NI:NF(K) = NF
170 BT = 2.302585093
180 PE =  INT ( LOG (E) / BT):UE =  LOG (E) / BT - PE
190 AE =  INT (1E5 *  EXP (BT * UE)) / 1E5
200 E(K) = AE:PE(K) = PE
210 POF =  INT ( LOG (F) / BT):UF =  LOG (F) / BT - POF
220 AF =  INT (1E4 *  EXP (BT * UF)) / 1E4
230 F(K) = AF:PF(K) = POF
240 PL =  INT ( LOG (L) / BT):UL =  LOG (L) / BT - PL
250 AL =  INT (1E3 *  EXP (BT * UL)) / 1E3
260 L(K) = AL:PL(K) = PL
270  FOR S = 1 TO K
280  PRINT NI(S);"  ";NF(S);" ";E(S);"E";PE(S);"    ";F(S);"E";PF(S);"   ";L(S);"E";PL(S)
290  NEXT S
300 K = K + 1
310  GOTO 80
```

A4
B5
C6
D7
E8
F9
G0
H1
I2
J3